CONSERVATION OF THREATENED PLANTS

NATO CONFERENCE SERIES

I   Ecology
II   Systems Science
III Human Factors
IV Marine Sciences
V   Air—Sea Interactions

I ECOLOGY

CONSERVATION OF THREATENED PLANTS

Edited by

J. B. Simmons, R. I. Beyer, P. E. Brandham, G. Ll. Lucas, and V. T. H. Parry

Royal Botanic Gardens
Kew, England

With a Foreword by **Sir Peter Scott**

Published in coordination with NATO Scientific Affairs Division by

PLENUM PRESS · NEW YORK AND LONDON

Library of Congress Cataloging in Publication Data

Main entry under title:

Conservation of threatened plants.

(Nato conference series: Series 1: ecology, vol. 1)
  "Proceedings of the conference on the functions of living plant collections in con-
servation and conservation-orientated research and public education, held at the
Royal Botanic Gardens, Kew, England, September 2—6, 1975, sponsored by the
NATO Special Program Panel on Eco-Sciences."
  Includes index.
  1. Plant conservation—Congresses. I. Simmons, John B. II. North Atlantic Treaty
Organization.
QK86.A1C66   639'.99   76-20762
ISBN 0-306-32801-1

Proceedings of the Conference on the Functions of Living Plant Collections in Conservation and Conservation-Orientated Research and Public
Education, held at the Royal Botanic Gardens, Kew, England, September
2—6, 1975, sponsored by the NATO Special Program Panel on Eco-Sciences

©1976 Plenum Press, New York
A Division of Plenum Publishing Corporation
227 West 17th Street, New York, N.Y. 10011

Printed in the United States of America

Organizing Committee

Chairman   Professor J Heslop-Harrison, Director
        Royal Botanic Gardens, Kew & Wakehurst, England

        Mr C D Brickell, Director
        Royal Horticultural Society's Garden, Wisley, England

        Professor K Esser, Director
        University of the Ruhr Botanic Garden, Bochum,
        Federal German Republic

        Professor J G Hawkes, Head of Dept. of Botany & Director
        University of Birmingham Botanic Garden, England

        Mr D Henderson, Regius Keeper
        Royal Botanic Garden, Edinburgh, Scotland

        Professor V Heywood, Head of Dept. of Botany & Director
        Plant Science Botanic Garden, University of Reading,
        England

        Mr G Ll Lucas, Conservation Unit
        Royal Botanic Gardens, Kew, England

        Dr S M Walters, Director
        University of Cambridge Botanic Garden, England

Conference Director   Mr J B Simmons, Curator, Living Collections
        Division, Royal Botanic Gardens, Kew &
        Wakehurst, England

Conference Organiser   Mr R I Beyer, Deputy Curator, Living
        Collections Division, Royal Botanic
        Gardens, Kew, England

Secretary to Committee Miss C Brighton, Jodrell Laboratory,
        Royal Botanic Gardens, Kew, England

v

Foreword

During the last two hundred years man has changed from living in equilibrium with the natural world which sustained him, to a new position in which he is now its undisputed ruler - and very often out of equilibrium - able in a matter of hours to reduce miles of forest to devastated, potential desert. This destructive and wasteful ability has increased dramatically over recent years. At the same time however the need for conservation, particularly of plants as a resource for the future, has also become apparent, along with the realisation that advanced technologies can produce more from existing agricultural and forest regions. This may to some extent relieve the heavy pressure on the vulnerable areas where short term over-exploitation leads to permanent destruction of whole ecosystems, and the attendant loss, for ever, of many of the animals and plants which originally lived there.

There still remains today a vast number of plant species whose potential is unknown. Maybe they will never have more than aesthetic value to mankind. But who knows where, for example, the next anti-cancer agent may be found. And anyway future generations may not be ready to accept such anthropocentric values, and the options should be kept open for the philosophical concept that all life on earth has a right to exist and that man has none to exterminate.

Nevertheless the destruction goes on apace, and we must rescue what we can, for a species is the end-product of 30 million centuries of evolution and therefore worth saving.

When a rain-forest stands in all its glory the individual trees often carry large numbers of epiphytes - orchids, ferns and bromeliads - and so when large forest areas are cleared, not only the tree species are lost but the epiphytes often have no natural habitat left and here the horticulturist has a major role to play. The Conference papers show strikingly how large numbers of species can be saved by the botanic garden and the horticulturist.

It gives me great pleasure, hope and even reassurance that this major group of bodies has recognised and responded so actively to this most necessary role in maintaining the riches of the plant world.

The threats are great and diverse but so is the potential of the botanic gardens and I cannot commend to them more strongly the Resolutions that were carried at this Conference, because without action they will remain mere words and hopes and this is not enough. The future lies in action now to safeguard the great diversity of plants for the future.

Sir Peter Scott, C.B.E., D.S.C.
Vice President and Chairman,
World Wildlife Fund
Chairman, Survival Service
Commission of I.U.C.N.

Preface

This volume contains the invited lectures presented at a
conference entitled "The Functions of Living Plant Collections in
Conservation and Conservation-Orientated Research and Public
Education" held at the Royal Botanic Gardens, Kew, from the
2nd to the 6th September 1975.

The Conference was attended by approximately 150 speakers and
delegates from twenty-eight countries of every continent. The
meeting was especially valuable in bringing together curators,
technical managers and members of the scientific staffs of many
European and North American botanic gardens who now look upon
conservation as one of their most significant functions.
Universities, research institutes and conservation agencies were
also well represented.

The Conference strongly asserted that the basic requirement
for the preservation of the threatened floras of the world was
conservation in the natural habitat by the extension of nature
protection policies and the setting up of an adequate network of
ecosystem reserves in all the major floristic regions. With this
as the fundamental principle, speakers examined the ways in which
living plant collections could be developed and managed as
conservation resources in their own right and in support of field
conservation through research and education.

The delegates quickly focussed on the serious problems now
besetting the flora of the world through the growth of human
populations and the demands of consumer societies. As noted by
Sir Otto Frankel (Canberra, Australia) in his address to the
delegates, ".... man has become the most powerful and destructive
organism the world has ever known", and much of his destructive
power is directed upon the plant kingdom.

Professor Raven (Missouri, USA) had stressed the special
problem of tropical floras. About two-thirds of the vascular
plant species of the world occur in the tropics, where the rain-
forest is now being felled or burned at a rate (according to the
latest FAO estimate) of 28,000 hectares a day. At the present

rate of destruction, Professor Raven estimates that at least 50,000 species will have reached threatened status or become extinct by the end of the century. Certain temperate floras are in equal hazard. Oceanic islands are particularly vulnerable, as in the case of the Hawaiian Islands, where as much as one half of the unique native flora is listed as endangered or now of doubtful status. The floras of regions with Mediterranean climates, where there now exist perhaps as many as 12,000 local endemic species, are also seriously at risk.

It was against this background that the functions of living plant collections maintained under human care were examined. A rough estimate suggests that throughout the world some 1,000 institutes maintain collections of vascular plants. With their wealth of resources, it is reasonable to expect that they could between them make a major contribution to conservation. In the case of the general botanic gardens, it was agreed that their most effective role would be in relation to their own floras, where a substantial contribution could be made to conservation in the field through scientific and technical services and also by maintaining resource collections in cultivation or in seed banks. Good examples representing this kind of effort may be seen in the work of the Canberra Botanic Garden in Australia, Kirstenbosch in South Africa, and Cambridge Botanic Garden in the U.K. with its special interest in East Anglian plants.

The view of many participants in the Conference was that with a proper co-ordination of effort between existing conservation organisations and institutions maintaining collections and seed banks much of the endangered temperate flora could well be saved. The problem is far greater for the tropics, and the prospects much less favourable. The Conference agreed that it was a responsibility of institutions in the temperate countries to give all possible help to organisations and institutions in the tropics in their formidable task, both in relation to conservation in the field and to the retention of samples of the vanishing plant "gene pools" in cultivation and in seed banks.

Various papers deal with the technical aspects of maintaining living collections. Professor K. Esser (Bochum, Germany) had noted that, strictly, it is possible in cultivated collections neither to conserve - that is, to keep populations under the same conditions as in nature following the same paths of evolution - nor to preserve - to keep them essentially in the same state as when they were taken in from the natural habitat. Nevertheless, notwithstanding the risks of genetic change in cultivated collections there may be no alternative but to retain species in this manner when extinction is their certain fate in the wild. The problems of assembling and maintaining living collections were also considered. To ensure that one at least begins with an adequate representation of the

genetic variation of each species one requires careful attention
to sampling methods, and to serve a proper function in conservation
and associated research, good documentation is essential. There
must also be careful consideration of what it might be reasonably
expected to achieve in each centre over the time scale envisaged
in any particular conservation programme. It is, for example,
impossible to ensure the survival of species in perpetuity when
cultivation depends on supplementary energy inputs, as in the
case of glasshouse-grown collections of tropical species in
temperate climates.

The problems of cultivating special groups are illustrated in
three of the papers by reference to mesembryanthemums in Hamburg,
Hawaiian endemics in the various Hawaiian Botanic Gardens and the
arctic flora of Greenland in Copenhagen. The costs of handling
ecologically specialised species are often very high, and can only
be justified in relation to research. The value of autecological
studies of plants in cultivation can nevertheless be considerable,
although there are often difficulties in transferring what is
learned to situations in the field. For the management of ecosystem reserves and for any future programmes of reconstruction
and rehabilitation of devastated ecosystems much more knowledge of
the reproductive biology of the species concerned is required, and
here again cultivated collections provide an important source of
research material.

Throughout the proceedings a great deal of stress was placed
on the need for better international collaboration and co-
ordination. For institutions to work effectively together information must be freely exchanged, and the need for standardised
recording and monitoring systems was noted by many speakers.
Improved data handling methods could simplify the task of locating
material held in cultivation and in rationalising the effort among
the different institutions. Examples of data systems of varying
degrees of elaboration, from standard card indexes to elaborate
computerised systems, were exhibited and discussed.

A major role for which botanic gardens in particular are well
adapted is in conveying information about conservation to the
general public. The millions of people who visit gardens provide
a ready audience, and given modern display and presentation
techniques the needs can be conveyed and sympathy and understanding
for the conservation case improved.

Much of the work of the Conference was conducted outside of
the conference hall, in smaller seminars, personal discussions and
during the visits to the gardens at Kew, Wisley and Cambridge.
The Resolutions that developed from these mutually stimulating
discussions can be seen as an assertion of the delegates' inten-
tions to use their new found international collaboration as the
starting point for active policies towards plant conservation.

To those who contributed to the success of this conference my sincere thanks go especially to the many able members of the Organising Committee and the Kew Organising Staff, who are individually mentioned in this volume, and also to the delegates and participants who combined through their support and contributions to give a real meaning and relevance to these proceedings.

My thanks are also due to Dr Andreas Rannestad, Secretary of the NATO Eco-Sciences Panel, who administered the NATO financial help that made this conference possible. Finally my thanks go to my co-editors for their help in producing this volume and to Mr D.V. Field for preparing the Index.

John B. Simmons
Kew, 1976

Contents

CONTENTS

Conservation-Orientated
Collections

INTRODUCTION

(Transcript)

J. Heslop-Harrison

Royal Botanic Gardens

Kew, Surrey, England

   This conference is being held at a time of mounting concern
for the future of the plant kingdom. Estimates based upon the
Red Data Book research and on surveys of threatened plants in
North America, Europe and other parts of the world suggest that
perhaps one-tenth of angiosperm species are in hazard of extinction,
a situation a good deal worse than might have been predicted even
ten years ago. Some floras are in greater peril than others, but
there are few major ecosystems not now at risk in some degree. The
cause is easily identified: natural events, including climatic
shifts, certainly account for some of the changes in species abundance and distribution now being observed, but the major pressures
arise from the manifold activities of Man.

   Much has been said in recent years about the stress we now
place on our environment, and for the present audience it will
surely be unnecessary to go over this ground again; but in the
context of this conference we would do well to recall that what we
have seen so far is likely to be eclipsed by what seems sure to
happen before the end of the century. The impact of Man on the
biome is already staggering in its magnitude, but it is certain
that some of the trends He has initiated will gather further
momentum, for they are - at this moment, anyway - well beyond His
control. The processes begun will continue; the environment
throughout much of the world will be degraded even more severely
than hitherto; natural ecosystems will continue to be eroded, and
as this happens more and more plant species will be driven to
extinction. As botanists we may deplore these facts, but I cannot
see that we have any alternative but to accept them. Given the
inevitability of change, our problem is to attempt to foresee the
outcome, and to consider what might be done to mitigate some of the
more baleful consequences.

There are, of course, many excellent reasons for urging a policy of conservation for plants, including the quite indisputable one that in the long run our own survival depends on a wise and balanced exploitation of the vegetation the world is capable of producing for us. The economic arguments can be supplemented by ethical ones, and I imagine there will be few plant conservationists who would dissent from the view that we have something of a moral duty to urge on the world policies that might give a chance of preserving for posterity some reasonable fraction of the richness and diversity of the plant kingdom as we ourselves have inherited it. Furthermore, there will be few who would contest that such policies must be based primarily on conservation in the field. The overriding task is somehow to ensure that adequately large and balanced samples of the principal ecosystems of the world are indeed set aside as reserves and managed so as to preserve the gene pools of all the associated species - plant, animal and micro-organism.- under conditions where the natural processes of variation and evolution can continue.

This then, is the background against which the deliberations of the conference must be set. Our purpose is to examine the part living plant collections can play in meeting the greater challenges of global conservation. The conference title is a cumbersome one, but it was the intention of the Planning Committee in adopting it to suggest that the potential functions are numerous, involving many facets of botanic garden activity. These functions are both direct and indirect - direct in the sense that living collections themselves can represent a valuable resource, and indirect in that they can offer important support for other conservation activities. Botanic gardens, for example, can perform a vital educational task in acquainting the public of the nature of human dependence on plants and in gaining sympathy and understanding for conservation policies. Plant collections have an essential role in support of physiological and ecological research which must be pursued to back up conservation activities in the field, and naturally they provide a valuable adjunct for the taxonomic studies required for the classification and cataloguing of floras.

Some of these matters are to be considered in detail in later papers, but without intruding too much on what is to come I would like to make a few preliminary comments. Firstly, there is the question of the value of living plant collections as conservation resources. This is a matter meriting the most careful attention from all of those managing living collections, and perhaps most of all for the older botanic gardens. It raises both philosophical and practical problems, and I may say that it has engaged a great deal of our attention here at Kew in recent years. The concept of garden collections as managed resources is no new one; after all, the original physic gardens were established as resource centres, and so were many fashioned in the tropics in the late eighteenth

century and through the nineteenth century, many under the tutelage
of Kew. But the sizes of the collections held in botanic gardens
have increased very greatly and so has the cost of maintaining
them; to justify the effort expended in upkeep and curation we
surely need a much better idea of what we are doing and why we are
doing it. Do we envisage conservation-orientated living collections
as repositories of rare and diminishing species, brought together
because of their impending demise whatever their economic, cultural,
amenity or scientific value? Do we look upon the commitment to
maintain such collections as an open-ended one? Do we envisage
the ultimate use to be made of the gene pools conserved in living
collections to be in the rehabilitation of natural ecosystems? Or
do we look upon living collections in the sense of museums, with no
other future before them but to be maintained in cultivation until
the taxa are lost through neglect, or until we tire of the
responsibility and deliberately abandon it?

In groping towards a policy for the very large living
collections held at Kew we have discussed some of these matters at
length, and have reached a few tentative conclusions. We do, for
example, accept that there must be priorities in maintaining and
building up conservation-orientated collections - based on local
interest, existing or potential economic value, scientific significance and other factors. We do see that for the activity to be
taken seriously at all it must be regarded as an indefinite
commitment - like curating a national library, for example, or to
offer a grimmer comparison, like maintaining a guard on radio-
active wastes from a nuclear power station. We do acknowledge
the collections as an important source for conservation-orientated
research and education. We do envisage them as being used as a
source for re-introduction into the wild; indeed this is already
happening with some of our holdings. At the same time, we do
recognise that for many species the only future is likely to be in
cultivation, and that the task of maintaining these will represent
little more than a sentimental burden not readily justifiable by
rational argument.

We also very clearly appreciate another point: that in taking
up the challenge of developing the living collections for their
manifold potential functions in conservation, botanic gardens can
find for themselves a new and purposeful role, at a time when
economic pressures are growing and their traditional functions
coming under question.

Secondly, I would like to comment on what we here at Kew
believe will certainly become clear during the conference, the
great desirability of collaboration and co-ordination in the future
between those maintaining living plant collections in the different
countries of the world. The tasks are immense; resources are
slender, and it is surely vital that they be deployed in the most

effective way possible. If we are to make an impact on the problems
agreement on broad aims will not be enough; there must be some
planning to achieve an efficient distribution of responsibilities
so that the various institutions are enabled to make their contributions where their climatic conditions and technical and scientific
expertise ensure the greatest effectiveness. I have in mind here
the possibility that particular establishments should accept special
responsibilities for different floras or taxonomic groups. Some
co-ordinating organisation will be required to optimise the effort
in this way, and we may note that two bodies already exist which
could play a central part - IABG, and the Threatened Plants Committee
of IUCN - to which I shall shortly refer. Collaboration of this
kind need not in any way intrude on the general operations of an
institution, and indeed it should be welcome in many centres as
giving a direction and meaning to existing specialist work.

Thirdly, we may note the very great importance now of communication, something to which this conference itself will surely
contribute. In matters of policy and practice, in horticultural
and in scientific matters, there must be good information flow.
Section IV is devoted to the vital topic of documentation, and I
shall not anticipate what this will cover. But I do wish to
emphasise the great need to improve our methods of disseminating
information about holdings throughout the world - not simply for
the benefit of other botanic gardens, but for conservation
organisations, universities and other potential users.

Beyond this there is the need for a better flow of information
about the technology of living collections - propagation methods,
growing techniques and other matters to be mentioned by Mr Shaw.
In my own contribution I hope to make a comment or two on a related
aspect: the need to make widely available phenological and other
data related to plant performance for use in field conservation
planning - for example, in devising strategies to permit regeneration, or even in re-introduction and establishment. Communication
in all of these matters is surely of the essence, for without it
our experience remains ours alone, however much it cost us to gain.

And, finally, may I refer to the role of the international
organisations. We are all conscious of the work of the UN in the
conservation field, and I refer here to the MAB programme, now in
its final phase, the burgeoning of UNEP programmes, and the continuing work of FAO. Equally important in the international scene
is IUCN, which has recently set up the International Threatened
Plants Committee. The work of this Committee will be described by
Mr Lucas, but foreshadowing his contribution a little I must tell
you that the regional and special group organisations of the TPC
are already being built up rapidly, and that we hope this conference
will provide a starting point for the third, an institutional network. The TPC can provide the required link between institutes

and the international agencies, and can clearly play a vital part in policy making and developing collaboration throughout the world. Mr Lucas will be appealing to everyone to participate in this work; may I endorse his words in advance.

THE GLOBAL PROBLEMS OF CONSERVATION AND THE POTENTIAL ROLE OF LIVING COLLECTIONS

G. Budowski

Director General, IUCN

Switzerland

INTRODUCTION

Generally speaking, one can distinguish three phases or periods in the process of developing a comprehensive programme for nature conservation, be it on a world scale, a regional or national level.   Conservation, as here defined, implies "the management of the resources of the environment - air, water, soil and living species including man - so as to achieve the highest sustainable quality of human life;   management in this context includes surveys, research, legislation, administration, preservation, utilisation and implies education and training".

The first period, which has lasted for thousands of years until the 20th century, is what may be labelled today, a lack of concern for conservation and one of its implications, namely, that of maintaining viable populations of living species of plants and animals throughout the world.

This is natural since it is only in recent times that there has been a considerable impact of mankind on resources usually of the exponential nature, with ever shorter doubling periods, usually between 10-30 years on a world scale depending on the resource affected.   This has led to the second period of leading to the awareness that there is a conservation "problem" through recognition of the finiteness of planet earth and the absurdity of projecting past exponential impacts on resource-use into the future.   Of course, this is not yet fully realised by a sizeable part of populations throughout the world, but no-one will deny that the trend is on.   Naturally, when reference is made to the lack of awareness over the past millenia, we are using a

terminology coined today and influenced by our present industrialise
civilisation.   Objectively speaking, we can assume that over that
period they were mostly small groups, often tribes, where an inheren
respect towards nature was part of the prevalent culture or, as we
would call it today part of "life style".

A third phase is concerted world efforts towards remedial
action but at this stage there is no clear indication that this is
forthcoming, at least at the scale requested in the light of the
staggering problems.   For some of the exponential impacts described
before, many different remedies - often conflicting among each other
- are being proposed.   No longer are past attitudes of "laissez-
faire" left unchallenged.   Or, to put it in other words, the over
simple assumption that "you cannot stop progress" is no longer
accepted in some countries.   Conflicts of interest between conservation and so-called development spring up in many aspects of life.

But all this does not lead to remedial action.   For many
people, any adaptation to changes be it material, psychological, or
ethical, has often been a more or less painful process and is
consciously or subconsciously opposed.   Moreover, conservationists
are far from being united in their approach for solutions and in the
conservation world there may be a whole spectrum of tendencies which
have often been compared with political parties which stretch from
far left - the radicals who advocate extreme solutions - to the far
right - the more conservative who propose rather mild changes that
can easily be channelled into our life styles.   And, of course,
there are many shades of "middle of the road" trends, a matter which
we in IUCN, representing some four hundred governmental and non-
governmental entities, are particularly well aware of.

For some people only certain political attitudes can make
conservation action prosper even if conservation problems are
relatively similar throughout the world regardless of the political
systems.   For others, conservation can only be practical once
present low levels of standards of life have been surmounted, even
if again, it is clear that conservation has been practiced in
cultures where material poverty has been evident for countless
generations.   This second period of awareness can then be labelled
as one of turmoil, of conflict, of claim for solutions, and,
unfortunately, also a lot of confusion.   But, fortunately, this
period has also pointed the way to some increasing action towards
solving apparent or real problems.   We see it in the increasing
number of national parks and reserves that are being decreed and more
or less managed, the species of plants and animals that are being
protected by law, the trade of endangered species that is being
curtailed, at least in some countries, and so many other achievements

Many of these actions, we should immediately admit, are usually based on an <u>ad</u> <u>hoc</u> decision and rarely reflect the results of careful planning related to a long-term strategy which should ultimately establish an acceptable balance between mankind and its natural environment.   Action is often the result of pressures - some kind of a "fire brigade" approach with many deficiencies and, perhaps more than anything else, with a certain disregard for obvious priorities.   Nevertheless, they show the intention and often successful action of the Governments and private groups throughout the world and they have a strong educational implication.   It is in this period that many actions even those that appear as premature, or lacking sufficient scientific background, <u>absolutely must be taken</u>, since any further delay could be catastrophic, leading for instance to the extinction of species.

Clearly our efforts in the next few years must take us to the third stage implying the promotion of conservation action based on the best scientific evidence, with a view to maintaining the natural diversity of the world, keeping options open for future action, and influencing policy and decision-makers in such a way that present and future development programmes are based on sound ecological principles with conservation aims receiving the high priority they deserve in our shrinking world.   On a world basis we are clearly in the first and second period and I hope that our meeting here will be regarded as a fundamental step to lead us to the third or well-balanced stage.

THE POTENTIAL ROLE OF LIVING COLLECTIONS

Living collections for conservation purposes are of course not a new concept, and I assume that everyone here holds the view that they can be extremely useful if based on a sense of purpose, backed by adequate scientific data and proper techniques.

Living collections have a long history both in plants (botanical gardens) and animals (zoological gardens).   In the work of IUCN we have in the past devoted more efforts to the role of zoological gardens in achieving conservation aims.   It is, of course, time we devote greater attention to plants.   I feel, however, that the experience accumulated with animals may be of great interest too, and I hope you will forgive me if I expand somewhat on the example of zoos, notably in breeding endangered animals that may relieve the pressure on wild species, the impact of education and awareness, and by generating more support for conservation action in the field. It has been the topic of many symposia and has also been a considerable cause of friction between conservationists and some zoos. The main criticism levelled has been that whilst scientific and educational values of zoos are well accepted, particularly for the better staffed and endowed zoos, the practical implementation for

conservation, namely the eventual release of zoo-bred animals into
habitats where they have been previously exterminated or greatly
reduced, is still to be witnessed on any scale.   In fact, zoological gardens have often been accused of being "consumers" of wild
species and the more rare the species becomes, the greater there
is a tendency for some zoos to pay large sums of money to acquire
specimens of such unique species.   Much more can be said about the
negative influences of zoos when seen from some conservationists'
viewpoint – as, of course, much can be said on the positive
influences.   If I have brought them to your attention, it was in
order to point out some of the dangers, or potential sources of
criticisms, which are likely to occur when it comes to living plant
collections.

Let us assume that a considerably larger number of botanical
gardens or other institutions will dedicate greater efforts towards
conserving endangered plant species by collecting seeds or viable
parts of the plant which lend themselves to a reproductive programme
under controlled conditions.   What will be the likely criticisms
to which such programmes will be exposed?   Let us quickly analyse
a few of them.

1. Collectors may pick up the last species, or considerably
   reduce the population of species from the last areas where
   they are found in the wild.   If not properly engineered,
   such action may generate considerable resentment from the
   people in the area, the country, or competing institutions.

2. Collecting of unique specimens from developing countries for
   the purpose of reproducing them elsewhere may be resented
   (remember Hevea from Brazil) and future expeditions may have
   entry permits refused.   There may be other negative effects
   on local conservation actions.

3. The genetic stock as reproduced in a botanical garden may
   differ considerably from what would be expected if natural
   evolution took place in the natural habitat.   There may be
   problems involved with an eventual reintroduction into the
   wild.

4. Introductions, if not properly controlled, could become
   pests or bring along pests with them.   There are other
   limitations that must be carefully evaluated before engaging
   in a fully fledged programme.

Now let us look at the positive sides of such action:-

1. Collections and propagations in botanical gardens can
   effectively save a species from extinction and in comparison
   to wild animals, for instance, it may be considered easier

to produce large populations of individuals of one species
in a relatively short time.

2. Well organised, such a collection can be used to re-establish
   endangered plants in their area and region once the danger
   that caused their depletion has been overcome.   Well
   engineered, such a programme could considerably help developing countries with endangered plant species in building
   up their own structures and programmes to effectively
   conserve and manage their unique resources for their own
   benefit and that of a concerned world.

It appears desirable, therefore, that in the various phases
dealing with the potential role of living collections, there is a
clear and unmistakable declaration of the conservation principles
which will promote the best long-term interests for conservation,
including the strengthening of conservation institutions throughout
the world, particularly those with meagre resources and operating
in regions where there is not yet adequate local support.   This
would make it clear that the intention is always to equally promote
the conservation of wild plants in situ, in their natural habitats
and that whenever possible management plans will be devised, and
assistance given whenever necessary to promote educational,
scientific, touristic (where compatible) aesthetic and other benefits
to those countries where endangered plants are collected.

Let us not forget, that conservation through living collections
is no substitute to conservation in the wild.   It is useful as a
complement, in some cases a fire brigade approach, but under no
circumstances should a breeding programme convey a soothing feeling
of reassurance, for instance, "now that we have good living collections, we need not worry anymore".

We need a code of ethics for those who collect live plants, and
a code of ethics for those who propagate them in controlled conditions.   Hopefully this meeting will examine these questions and make
it unmistakably clear that this is a stepping stone towards a better
world where we can transmit to future generations the rich and
diverse heritage bestowed upon us by millions of years of evolution.

DISCUSSION SECTION I CONSERVATION-ORIENTATED COLLECTIONS

Mr Henderson (Edinburgh) asked whether there was any way of avoiding the complacency that Dr Budowski mentioned could occur when all threatened species are brought into living collections. Governments might feel that cultivation in a botanic garden is a cheap alternative to conservation in situ.

Dr Budowski (IUCN Switzerland) replied that part of the botanic garden's programme and aims should be involved in working for conservation action by promoting the establishment of reserves, training local people, etc. This would make it clear that the primary interest of the garden was not to enrich its own collections but that it was concerned with practical conservation problems.

Sir Otto Frankel (Canberra) raised the important distinction between maintaining genetic diversity within and between species. This difference often has been blurred as in the UN Stockholm Conference (1972) so as to engender a general appreciation of the need to salvage genetic diversity in living collections. Conservation only at the species level, is usually undertaken in living collections. Sir Otto stressed that collections should make great efforts to solve the more difficult problem of maintaining intra-specific diversity and suggested it as an important matter for discussion.

Existing Collections

PAST: AIMS AND OBJECTIVES

B. F. Bruinsma

Leiden Botanic Garden

Netherlands

INTRODUCTION

Before I start to say something about the functions and aims of collections of plants in the past, I should point out that almost all of the oldest and remaining collections are maintained by universities and related teaching institutions. So, if I tend to refer to the situation in the Netherlands and in Leiden in particular, I must beg you not to interpret this as pure chauvinism but more as a historical necessity and of course, as a result of the fact that my work is in Leiden.

However modern the present structure of botanic gardens may seem, it actually represents adjustments to the traditional con-figuration of the old concept of the Hortus, and this is one of the reasons why the staffs of botanic gardens are beset by so many problems at present.

THE HORTUS MEDICUS

The oldest botanic gardens - and these were the ones with a function in higher education - were highly restricted; they played a part in the teaching of medicine, since the students were required to recognize the medicinal plants and to know something about their morphology and application. The place of the taxon in the plant world was of no importance. In the course of the centuries, however, the botanic gardens could hardly escape the influence of scientific developments, and eventually they were put to a much broader 'use'.

Before the establishment of the first Botanic Garden (the
very first being that of the Vatican, called the Orto Vaticano,
dating from the first half of the sixteenth century and followed
by the later Italian gardens), the available knowledge about plants
was based mainly on Dioscorides (who lived in the first century of
the Christian era) and Galen (second century) as well as the plant
lore of the Arabs. After the Middle Ages, during which the writings
of Dioscorides and Galen were accepted unquestioningly, a critical
attitude toward these authors developed. Until this occurred,
religious thinking had led to the assumption that all plants grew
everywhere, and little was known about geographic botany or plant
societies, but after medieval times there was an inclination to
observe the plants themselves.

With the growth of a more critical attitude, individual
scientists began to have considerable influence. Four well-known
botanists - Fuchs, Dodonaeus, Clusius and Matthioli - saw geographical distribution quite clearly (Clusius's _Flora pannonica_
provides an example), and at the same time the ornamental value of
the plants began to play a part in the composition of collections
even though the medical importance still predominated. In addition,
at the end of the sixteenth century, the first steps were taken
towards taxonomic teaching, mainly in France, Italy and the
Netherlands but later in Great Britain as well. One of the first
to take up this field was Caesalpino.

As more knowledge was gained, the composition of the
collections in the earlier gardens changed; they no longer consisted solely of medicinal plants but evolved into a kind of hybrid
representing both the ancient interests and the results of the new
criticism. The new pattern showed a large core of long-known plants
around which newly imported plants were arranged. Much time and
money was spent on these collections, and they led to comparison of
the 'new' with the 'old' plants as well as to the display of not
only the species with therapeutic value but also other species of
the same genus, which it was pointed out must not be used for
certain prescriptions. This comparison led in turn to the first
faint approach to a plant taxonomy. In these pre-Linnean times
plants were compared at the genus level rather than by species,
and they were arranged in the gardens according to genera. Meanwhile, the importance of the plants for medical science decreased,
even more strongly in the second half of the seventeenth century
when chemistry began to acquire scientific stature and chemical
prescriptions came into medical practice.

But scientific 'curiosity' was not the only motive for information about plants to be obtained. Economic factors played an
important role too.

VOYAGES OF DISCOVERY AND TECHNICAL DEVELOPMENTS

As early as the late Middle Ages, dried 'parts' of plants, the
so-called simplicia, were imported from the Orient, for instance
cinnamon. This trade was conducted via Venice, but the wars between
Venice and the Turks in the sixteenth century caused it to stagnate
and the European physicians saw their supplies dwindling. At this
time it was generally accepted that the plants of Western Europe
were less suitable for medicinal purposes, and they were in much
less demand. The decline of the Venetian trade reached its lowest
point around 1550, but in the meantime the Portuguese had settled
in India and a more reliable trade in spices and simplicia
developed via Lisbon. Contemporary books show which plant parts
were traded in and what prices were paid for them. As a result of
this situation, too, Portuguese and Spanish physicians came into
contact with those who supplied their herbs and spices, and learned
more about the appearance of the plants whose products they had
used so long. This was also why Garcia da Orta (whose name, by
coincidence means 'of the garden') could criticize such authors as
Dioscorides on the basis of his own observations. He was then
living in Goa; furthermore, his criticism was more soundly based
than that of Fuchs and Dodonaeus. It was in this period, too,
that the first attempts were made to cultivate plants from the
Far East in the West. Unfortunately, many of these attempts failed
because such technical provisions as reliable glasshouses and
especially stoves were not available.

The middle of the sixteenth century saw an important
'invention' however; this was the Herbarium. Books containing
dried plants gave an enormous impetus to the acquisition of knowledge about plants, and it is known that directors of the Leiden
Hortus Botanicus, among them Pauw and Clusius, persuaded voyagers
to the East Indies to bring back dried plant material for them to
study, and sometimes even sent someone along on such voyages
especially to collect such material, which was also used for
teaching purposes.

In some cases it was the ship's doctor himself who - like
P. Hermann and Rumphius - hunted for plants. Five volumes of
Hermann's celebrated herbarium (collected in Sri Lanka during his
employment by the Dutch East India Company) are still preserved in
the British Museum for Natural History, and part of Rauwolf's
herbarium is to be found in the Leiden Herbarium.

When technical progress made it possible to build greenhouses
and invent the stove for heating purposes, it became increasingly
easier to create the right conditions for growing tropical plants.
Hothouses were then extremely expensive, and were first put to use
in the university garden of Leiden and the city garden of Amsterdam
at the end of the seventeenth century, when these institutions were

quite wealthy.   Some lengths of tile conduit still remain in the
soil of the Amsterdam Botanic Garden and are remnants of a very
early heating system.   In England the system was soon adopted by
the contemporary Duchess of Beaufort, who perfected it, and also
attracted the interest of the Fellows of the Royal Society.   All
these technical provisions made it possible to show medical
students foreign plants, and a good deal of effort was made, in
particular by Hermann, Commelin, and later by Boerhaave to fill
the heated glasshouses with such material.

TAXONOMY

From 1680 on, Hermann held a professorship in the University
of Leiden.   He developed a serious interest in plant taxonomy, and
may be said to have been the first to arrange a taxonomic lay-out
of the garden and to teach the plant system of John Ray in it.
The plants in such gardens were provided with numbers, and these
numbers correspond with registers.   In this period a professor who
could publish a well-annotated plant register felt that it marked
a high point in his career.

Together with France and its Jardin des Plantes, the
Netherlands found themselves in the vanguard of botany, led by
Tournefort in Paris and Hermann in Leiden.   Hermann was presented
to Louis the Fourteenth at Versailles as a princeps botanici, a
king among botanists, which led the king to inquire why the
fountains of Versailles had not been turned on as befitted such a
'royal' visit (a very costly gesture).   Dutch sobriety did not
tend to such epithets, although it was later used for Linnaeus.

The enormous expansion of plant taxonomy and the glorious rise
of zoology in the nineteenth century caused biology to be separated
from medicine about 1870.   A man like W.F.R. Suringar, the
sixteenth director of the Hortus, had had a medical training but
became the country's first professor of biology.   At present, this
separation is sometimes a matter of mild regret;   medical students
no longer come to the garden to learn about plants (with the
exception of an occasional student with homoeopathic inclinations),
although the pharmacology students still come to study plant
collections at the beginning of their training.

PRIVATE COLLECTIONS AND PLANT HUNTING, THE GROWERS (NURSERYMAN)

The introduction of private plant collecting was closely
related to the expanding trade of the seventeenth and eighteenth
centuries.   The directors of the Dutch East and West India Companies
resided 'at the source', which also explains why more plant material
flowed from private gardens to university gardens than in the

reverse direction. The wealthy owners of private gardens regularly
made gifts of plants and other objects to enrich the university
collections. There are few records of gifts from university
collections to private gardens, but it is known that the chief
gardeners were officially forbidden to trade in plants. It was
not until the nineteenth century that plants were sold to private
persons on a modest scale in the Utrecht Botanic Garden.

At that time it was much more difficult than it is now for
teaching gardens to specialize in a particular field. (Current
examples are provided by Leiden's specialization in Asian, Utrecht's
in American and Wageningen's in African plants).

In the earlier centuries specialization was mainly restricted
to the private collections. The garden of Simon van Beaumont (who
was a friend of Hermann) had a collection of plants from what is
now Surinam, and if Boerhaave is regarded as a private individual
rather than in his capacity of director of the Leiden Hortus, his
own garden on his Oud Poelgeest estate may be seen as Arboretum,
the only survivor of which unfortunately enough, is the remnant of
a large <u>Liriodendron tulipifera</u> he planted himself.

Nevertheless, due to indirect economic factors, there was a
slight tendency toward specialization in the formation of
collections, including those of the Botanic Gardens, as early as
the seventeenth century. This came about as a result of the
expansion of Dutch power and trade, which meant that mainly Asiatic
plants reached the gardens at home. The same held for England with
respect to North America (Virginia), and France received such
plants as <u>Robinia pseudoacacia</u> from Canada.

Because Germany was not a naval power at this time and had no
colonies, its gardens lacked foreign plant collections and were
dependent on gifts from Holland, England and France.

One aspect of botanic gardens which flourished in Britain and
America but was hardly felt in the Netherlands apart from
von Siebold's work, was the custom of rich amateurs to send
expeditions out to collect wild plants. In Great Britain this
plant hunting was supported by wealthy members of the upper class,
who financed voyages which added valuable plant material to private
gardens (like that of the Duke of Bedford), but also supplied
botanic gardens (like the Hookers in Kew) and commercial nurseries
(such as James Veitch & Sons). The hunting of plants in distant
regions first occurred in fifteenth century Italy.

There was, however, another factor besides plant hunting which
gave a strong impetus to the development of the botanic gardens in
the nineteenth century, and this was the rise of socialism. In
England in particular, physicians put a strong pressure on the

government with respect to care for paupers. In the Netherlands,
public lectures on agriculture were given and in Utrecht in particular on horticulture. Around 1820, there was a strong feeling
that something must be done for 'the common people', and these
efforts led to an improvement in the nutrition of the poor. In the
botanic collections of this period, parts of the garden were
reserved for edible plants and fruits. This was the case in the
Utrecht Botanic Garden but not in Leiden; in Utrecht too, the
classical secondary schools offered a choice among seven fields of
training, one of which prepared the students to become a
pharmacist's assistant. In our time the collections in botanic
gardens are regularly used for teaching and/or research in
horticulture at all educational levels.

BOTANY: RESEARCH

Starting between 1720 and 1730, experimental botany began to
gain ground in Great Britain. This field of research was less
dependent on teaching gardens, since it used plants which could
easily be grown indoors and in ordinary gardens. Unusual plants
and large quantities were not needed.

But once new methods of research were introduced and experiments had to be repeated many times, the demand for plants
increased and with it the need for more space.

EDUCATION AND RECREATION

Botanic gardens were open to the public as early as the
sixteenth century, but only by appointment and under the guidance
of the hortulanus. (The word publicus in the term Hortus Publicus
is referring to the fact that such gardens were financed by the
government and not to the admission of the public, in other words
with 'public funds').

Visits to a botanic garden may well be called an early form
of recreation, especially in view of the important role these
gardens often played in the education of sons of upper class
families. It will be recalled that in this period it was fashionable
for the fathers of these young men to send them on a grand tour,
which whenever possible included visits to famous botanic gardens
in other countries. Nevertheless, old travel books and diaries
show us that even then plant collections were considered rather an
odd idea, and although these young men were to hold high positions
in society later on, hardly one of them ever concerned himself
with such collections. The world of botany was seen as inhabited
by a small, highly selected circle, and it is remarkable to see
how much this relatively small group actually accomplished.

Many stories about the gardens have come to us. It seems for instance that the Burgomasters of Leiden had a key to the Hortus, to which they retired after important meetings for further, less official discussions, and Herman Boerhaave is known to have invited his students to converse in the afternoon at Oud Poelgeest, where Linnaeus joined them during his Leiden period. On these occasions the company did not indulge in 'shop talk'; they made music (Boerhaave himself played the flute), recited poems and discussed botany.

In the Golden Age and in the eighteenth century the private gardens were important status symbols. Political and business affairs were often discussed in them and in the glasshouses and the menageries. They reflected the wealth of the owner and of the country, and provided an agreeable setting for a relaxed conversation. The records show, for instance, that in 1690 Burgomaster Witsen of Amersterdam personally conducted the French Ambassador on a tour of the Amsterdam Botanic Garden, showing him the Dutch domination of the flora of four continents and consequently the worldwide power of the Dutch Republic.

It is known too, that certain problems encountered in the cultivation of plants from the Far East were solved with the help of information provided by the Dutch East India Company. There had been difficulties with watering, the exact provenance of the plants was required, and so on. In some cases questionnaires were even sent to colonial Governors to obtain details concerning plants.

For private collections, great variety was not considered very important. It was not until the nineteenth century that more attention was given to diversity.

THE INFLUENCE OF BOTANIC GARDENS ON ECONOMICALLY IMPORTANT PLANTS

Botanic gardens started to play a role in the establishment of commercial plantations in the seventeenth century, when the Dutch East India Company wished to know, for instance, whether the cultivation of the nutmeg could be limited to a given region (in order to form a monopoly); was there a botanic barrier? Germination proved difficult to induce, even in botanic gardens, so that the theft of seed for cultivation elsewhere generally failed.

In the seventeenth century Hermann made a thorough study of cinnamon cultivation in Sri Lanka, but when Sloane (of Chelsea Physic Garden) - with whom he corresponded at length - asked for information about growing this crop, Hermann found Dutch economic interests too important to supply him with any of the details.

The Amsterdam Botanic Garden sent material of the coffee plant

to Java in the hope that a coffee monopoly could be established in
this Dutch colony. However the attempted monopoly failed, because
Louis the Fourteenth had also received material and arranged for
its cultivation.

RECENT DEVELOPMENTS AND PROBLEMS

Today, the botanic gardens are no longer closed institutions.
Old elements such as scientific contacts and the exchange of seed
persist, but their function in university teaching and research
has been largely displaced by experimental work. The varied
collections require a large amount of labour and cost relatively
large amounts of money to maintain. In the Netherlands the
situation is complicated by the prevailing economic conditions,
but on the other hand public interest in plants has increased
enormously and there is also a growing interest in the educational
aspects at the sub-university level.

Various approaches are being applied to achieve specialization
in the collections and also to maintain them as efficiently as
possible, but it is difficult to train professionals and there
seems to be less and less interest in and commitment to this work.
In the Netherlands there has even been talk of merging some of the
gardens, for instance the two in Amsterdam with the one in Leiden.

THE FUTURE

The fate of botanic gardens remains to be seen. In the meantime, the staffs of the gardens in the Netherlands are exploring
ways in which the functions and objectives of the collections
could be usefully adapted to ensure their continued existence.

PRESENT: THE RESOURCE POTENTIAL OF EXISTING LIVING PLANT COLLECTIONS

John B. Simmons

Royal Botanic Gardens

Kew, England

INTRODUCTION

In principle, it is generally agreed that the creation of ecosystem reserves offers the best means of conserving threatened floral regions, but unfortunately in practice this does not at present reduce the need for the preservation and rescue of a great number of plant species with natural world populations that have declined to an inadequate survival level and for which properly protected reserves will come too late.

To help solve these problems the existing international network of botanic gardens can play and is now, in part, playing an increasingly valuable role by developing public awareness in plant conservation, creating and managing ecosystem reserves, establishing propagating and distributing endangered plants, and forming seed banks. Equally important is the creation of facilities for fully documented autecological studies on threatened plants. For such is the scale of the problem that we neither know what we are losing nor even its potential value.

But the needs of conservation are more scientifically exacting than most existing commitments. To meet these needs we must look again at the way we manage our collections.

REPRESENTATION OF TAXA

Most botanic gardens are concerned only with a fraction of the extant phanerogams and pteridophytes; many families and most of the lower orders of the plant kingdom, such as the _Algae_ (sea-

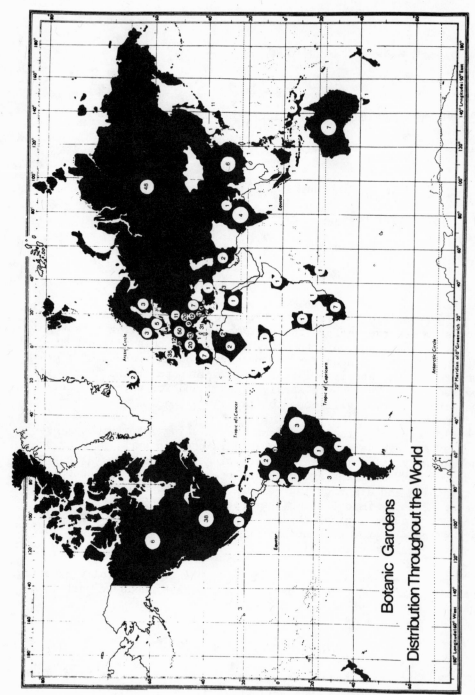

Botanic Gardens
— Distribution Throughout the World —

Figure 1

weeds for example) are largely ignored. At Kew 40% of the flowering
plant families are not represented in the collections which are thus
imbalanced, similarly the total resources of botanic gardens hold
only a relatively small proportion of the world's flora. Taking Kew
as the example again, had not botanist and horticulturist inter-
vened, the area of Kew might now be fields with as few as 300-400
native British plants, but by the application of human skills we
can in fact hold some 80,000 separate accessions in a little over
300 acres (124.4 hectares). Accepting then the limitations of the
survival capacity of a species in cultivation, it is now reasonable
for individual gardens to consider improving the use of their
resources by developing specialist international collections which
can in turn lead to a rationalisation between botanic gardens.
Together these botanical collections could then encompass a very
high proportion of the present estimate of 20,000-25,000 endangered
species.

The world map of the spread and number of botanic gardens
(Figure 1) shows that they could easily become an effective network
for conservation purposes. It also shows the high density of
gardens in the Northern Hemisphere and the comparative lack of
gardens in the tropics. The holding potential of an individual
garden can be seen by reference to the numerical development of
the collections at Kew which over its history have increased in a
broadly linear pattern with 3,389 "species" recorded in 1768;
11,013 in 1810/15; 20,550 in 1899 and 25,000 in 1931/34.

The end of the nineteenth century was a turning point for the
collections at Kew for it marked the end of the major expansion
of glasshouse accommodation (most of the subsequent changes have
been redevelopments) which led to the numerical levelling out of
the tender plant collections. But it also signalled the beginning
of the greatest phase of discovery and expansion for hardy plants
including herbs but particularly the tree and shrub collections
which increased from just over 3,000 species and varieties in 1899
to over 7,000 in 1931/34. However, the higher numbers of the late
nineteen-thirties were almost certainly achieved by the inclusion
of cultivars (described then as varieties); current figures do
not include cultivars.

During the Second World War the collections were depressed
and post-war years saw only a slow recovery until 1965 when the
addition of Wakehurst Place provided an opportunity for the expansion of the tree and shrub collections and added a conservation
potential for regional and homoclime flora extension.

The Kew collections are once again expanding but now with
increased scientific objectives.

SURVIVAL CAPACITY

The ability of a species to survive in cultivation has in the past contributed largely to its representation within botanic collections. This survival capacity has its base in the environmental tolerance, regeneration and wound healing potential of a species.

Cultivation tolerance is mostly seen to have a taxonomic relationship but with endemic species it is not unusual to find that paleoendemics are more difficult to cultivate than neoendemics.

It is important to analyse collections in this way eg. cycads (with the exception of Bowenia) have both a wide environmental tolerance and good wound healing capacity. They are therefore easy subjects for cultivation, and can be maintained over centuries. In contrast the endemic Tasmanian montane ewartias have a narrow environmental requirement and are difficult to grow.

Thus plants that are common in cultivation normally possess advantageous cultural characteristics. For example, Australian Eucalyptus species are noted for their good regeneration from seed, and seed of course has been the basis of international botanical plant exchange; Kew's first seed list was distributed in 1885.

The New World bromeliads are also successfully cultivated since they have a good regeneration potential and their xerophytic habit contributes to their wide environmental tolerance. Within one fairly light but humid glasshouse with a base minimum air temperature of $60^{\circ}F/16^{\circ}C$ it is possible to grow plants from the high Andes such as the puyas; desert plants like the dykias; the widely temperature ranging Tillandsia usneoides; Amazonian forest billbergias and tropical montane species such as Brocchinia reducta from Mount Roraima. While this may not be an ideal situation it does show that the plants have a remarkable range of environmental tolerance and is particularly significant when one recalls that some of the genotypes have been in continuous cultivation at Kew for over one hundred years. Our existing specimen of Hechtia argentea was actually received at Kew about 1870 or earlier. Conversely it is intolerance to cultivation which has limited the range of plants that are actually grown and so helped to define what may be described as the international botanic garden flora.

By means of such 'survival' evaluation one can determine the groups that have a priority need for ecosystem reserves and those for which cultivation could be an alternative.

But gardens should aim to provide more than survival conditions, and to meet conservation needs, optimum growth, flowering and seeding should ensue. Flowering is critical in another aspect;

the plant must be taxonomically verified. For example, seeds which
were supposed to be from the rare endemic Lebronnecia kokioides were
collected from its one remaining location in the Marquesas Islands
(S. Pacific), but after germination they proved to be a Gossypium!
Fortunately Lebronnecia had already been established in cultivation
by the Pacific Tropical Botanic Garden in Hawaii and at Kew as
seedlings from the Hawaiian plants.

USE OF ENVIRONMENTS

Within gardens satisfactory environments must be provided for
the collections and for successful conservation the environments
provided by a garden must be secure in perpetuity. Technically we
now have the capability of providing a great range of environments
but dependence upon applied energy sources, whether as management
care or as fuel, reduces security. The favoured environments for
conservation are those which are both cost-effective and in need
of the minimum management input. Using Kew as an example its
living collection environments can be classified as follows (see
also Table of Environments at Kew, Figure 2):

1. Natural. In Kew and Wakehurst we have two Northern Cool
   Temperate Zone gardens. Kew offers a 20 m. deep acid sand
   river terrace overlying clay with a low rainfall, average
   635 mm. per annum, and Wakehurst a wealden ridge with a
   greater diversity of acid sands and clays overlying sand-
   stone, a variety of aspects, an altitude of 134 m. above
   sea level and an average of 821 mm. per annum rainfall.
   These natural conditions can be managed in one of two ways:

   a. Maintenance of existing native plants by ensuring areas
      are not disturbed through tree removal or drainage, and
      by mowing grass areas only once a year in late summer
      or winter, depending on plants involved. For example
      at Kew the natural grass areas hold Goldilocks
      (Ranunculus auricomus) and Meadow Saxifrage
      (Saxifraga granulata) and the dark shade of yew trees
      (Taxus baccata) at Wakehurst forms a refuge for the
      Tunbridge Wells filmy fern (Hymenophyllum tunbrigense).

   b. Introduced plants that regenerate naturally which may be
      of local or other homoclime origin can in some instances
      become weeds and need some control eg. Montia (Claytonia)
      perfoliata and the Giant Hogweed (Heracleum mantegazzianum).
      Others may be encouraged by late grass cutting etc. to
      maintain their population, eg. at Kew Smyrnium perfoliatum
      (S. Europe to Asia Minor) in the Queen's Cottage Grounds
      is encouraged in this way.

LIVING COLLECTION ENVIRONMENTS PROVIDED AT KEW

Description	Labour	Input Materials	Fuel	Security Status	Comments
Natural					
a) Existing flora	v. low	v. low	v. low	v. high	Few taxa covered
b) Introduced plants	low	v. low	v. low	high	Can become weeds
Management Dependent					
a) Modified local habitat	medium	high	low	medium	Trees more secure than herbs
b) Competition controlled	medium	low	high	medium	Requires trained gardeners
c) Regeneration control	high	high	medium	low-medium	Requires skilled labour and management
Fuel Dependent					
a) Glasshouses - cool, ambient	v. high	v. high	medium	low	Extends range of work
b) Glasshouses - hot, moist	v. high	v. high	v. high	v. low	Holds intensive collections
c) Controlled environment equipment	v. high	v. high	v. high	v. low	Used for experimental work

Figure 2

2. Management-dependent collections. In this category the natural climax is used as the base which is modified in various ways, for instance controlling natural competition and regeneration so as to extend the range of biotopes and thus increase the density of the collections. These factors increase both cost and risk.

   a. Modifying the local habitat by, for example, changing soil composition through the addition of fertilizers and organic materials or changing pH by addition of lime for calcicoles or sulphur for calcifuges. Water may be used to provide a moister micro-climate or to create features, such as ponds or bogs.

   b. Competition control by removal of naturally competing plants through mulching, mowing, weeding, tree thinning, etc. In this context rock gardens are more labour intensive than arboreta.

   c. Regeneration cycle. The form of life cycle is of con-servation significance. Trees such as the beech (Fagus sylvatica) with an expected lifespan of 200 years at Kew are obviously a more suitable proposition for a living collection than annuals which are better adapted for seed banking. Seed is both the natural and most economic means of propagation but because of problems with genetic purity, particularly in botanic gardens where open pollination occurs, vegetative propagation may be favoured. Open ground division of plants is probably the most economic means of vegetative propagation with grafting the most expensive. However, in conserva-tion terms cost is perhaps less significant than that of ensuring an accurate follow through of recording at the propagation stage when the cuttings of one taxon look much like the cuttings of another.

3. Fuel-dependent environments. These are primarily the glasshouse collections. Glasshouses are reasonably energy efficient structures which make the maximum use of natural sunlight, but need supplementing with heat, moisture and growing medium. Further additions such as supplementary lighting, carbon dioxide injection and artificial cooling can be used the problem is they are all now very expensive.

   The warmer and moister the climate required, the more the energy needed and despite automation, glasshouse collections are still labour intensive. Also the constraint of height has biased collections towards herbaceous plants and led to a poor representation of tropical dicotyledonous trees. However, in their favour numerically large collections of

taxa can be held in small areas eg. 1,500 taxa (2,637
accessions) of ferns at Kew in just over 1,000 square metres
of growing space;   an average of three accessions per square
metre.

The development of the <u>reserve collection</u> concept can also
aid conservation use of glasshouses. At Kew many of the
display houses have been "landscaped" both to improve the
settings for the plants and to reduce labour input. All of
the display material is expendable in scientific terms since
reserve stock and rare plants are held behind the scenes in
automated glasshouses along with the propagation material.

IMPLICATIONS

Further criteria emerge from this analysis, and it can be
easily seen that there is a very great advantage in a botanic
garden working on the native flora of its region, for example,
Kirstenbosch in South Africa with its regional gardens and Canberra
in Australia with its native plant garden, and there is everything
to be said for botanic gardens establishing and managing natural
reserves where this is appropriate and no other active agency
exists.

The management of natural reserves usually falls down with
inadequate policing;   I have seen the hunters' dogs and heard gun-
shots echoing through tropical reserves. This problem occurs
throughout the world where local populations cannot afford to or
have not been educated to care about their native animals, and
(especially in forest practice) care still less about their native
plants. A great deal depends upon the local economic and political
climate but we only need to observe some of the present uncontrolled
practices of British agriculture eg. wholesale hedge and pond
removal, to realise that this problem is close to us all.

Since most botanic gardens have to serve other needs,
collection extension away from regional plants should favour plants
from homoclimes (parallel environments in other regions) since
such plants are likely to form effective propagation groups and to
be the most secure and economic in conservation and cost terms.
Beyond this it is reasonable for well-endowed gardens to assist
in regions of the world that lack the expertise and resources to
safeguard their own floras.

For Red Data Book plants we have, for example, already
developed a pilot co-operative scheme whereby for example
Adelaide Botanic Garden has accepted Somali plants that could not
be housed at Kew.

In summary it may be concluded that collections that depend upon man-made energy sources for their survival should not be the basis for conservation work except where no more secure natural environment exists.

USE OF COLLECTIONS
(with Conservation in mind)

1. Propagation. Most plants, unlike the higher animals, have the capacity for vegetative propagation. Also many plants, for example the species orchids and even monoecious palms such as the giant Chilean honey palm (Jubaea spectabilis) prove self-fertile and thus viable seed can be produced from a single plant. With this multiplication advantage over animals advanced propagation techniques can be used with great effect to aid conservation.

   Clonal propagation of advantageous genotypes has been an important factor in the improvement of cultivated plants. Kew, for example, has contributed significantly to the selection and propagation of hardy ecotypes for introduction to Temperate Zone gardens. The disadvantages are, however, the susceptibility of clones to epidemic pests and diseases and the possible build up of systemic pathogens (viruses and bacteria) through poor horticultural practice.

2. Distribution. Through propagation stock becomes available for distribution and use. The distribution of plants (as vegetative propagales or seed) from Kew still forms an important part of our work and every week consignments leave for destinations all over the world. Less emphasis is now placed on reciprocal plant exchange - the benefits come in many other ways - but bona fide workers in other institutes request and receive plants of medicinal, breeding, other scientific, economic, amenity or educational value. In 1974, 315 consignments of plants (excluding seeds) containing 1,485 taxa (c. 16% of the collections) were sent to twenty-seven countries, (Figure 3) and these figures have increased in 1975.

   Recent examples continue to show the valuable range of uses eg. Medicago spp. to the Argentine for a fodder breeding programme, Hieracium to Hamburg for land erosion studies, Euphorbia coerulescens to the National Blood Transfusion Service whose initial research scan of plants at Kew has shown that this species contains substances of possible value with work on blood grouping lectins.

   Even our "rubbish" has value, recently the World Health

ROYAL BOTANIC GARDENS KEW
DISTRIBUTION OF PLANT MATERIAL 1974

RECIPIENT COUNTRIES	TOTALS		ANALYSES OF DISTRIBUTIONS									
			BY OFFER FROM KEW				REQUESTS BY OUTSIDE ORGANISATIONS					
			Conservation material		Surplus material		Research projects		Amenity use etc.			
									educational		ornamental	
	batches	taxa	batches	taxa	batches	taxa	batches	taxa	batches	taxa	batches	taxa
Argentina	1	7					1	7				
Australia	6	38			2	24	2	2			2	12
Canada	6	28					5	27	1	1		
China	2	15							1	3	1	12
Denmark	4	60			2	57	2	3				
France	6	34	2	2	1	1	2	13			1	18
Germany	18	126	6	8	7	96	2	4	2	2	1	16
Ghana	3	3					3	3				
India	10	41					9	39			1	2
Israel	1	19							1	19		
Italy	2	26					2	26				
Ivory Coast	1	44									1	44
Jamaica	1	1									1	1
Japan	3	16					1	7			2	9
Kenya	1	1					1	1				
Madeira	1	16									1	16
Malaysia	14	21					14	21				
Netherlands	7	28					2	3	1	1	4	24
Norway	1	2					1	2				
Rhodesia	1	2					1	2				
Singapore	3	5			1	2			2	3		
South Africa	5	41	2	3	1	36			1	1	1	1
Sweden	2	23			1	15			1	8		
Switzerland	2	4					2	4				
Tahiti	1	2							1	2		
United Kingdom	187	843	15	39	38	335	76	207	21	75	37	187
United States	24	148	1	1	2	33	12	63	4	44	5	7
U S S R	1	1					1	1				

Figure 3

Organisation took some old giant bamboo stems to use in evaluation studies on new insecticide formulations for use on house wall panels.

The distribution of rare or endangered plant species is increasing, some on receipt at Kew (if the quantity is sufficient) and others following propagation at Kew. In 1974 some forty-three taxa were distributed as plants in this way. The recipients are botanic gardens indicated a willingness to give the plants a secure home on the premise that establishing a species in several centres increases its chances of survival in cultivation. Examples include the propagation and distribution of Cestrum psittacinum which has been in cultivation at Kew for many years but is now thought to be extinct in the wild. Other material distributed on receipt at Kew has, for example, recently included from Mauritius, Foetida mauritiana, Latania loddigesii and Hyophorbe ver-schaffeltii and from Aldabra Lomatophyllum aldabrense.

3. Recording. By improved observation, experiment and recording of living plants valuable information can be built up for analysis and publication. For the majority of rare plants there is little or no information available on their environmental needs, tolerance to cultivation, generation cycle, time to first flowering, whether or not self-fertile, means of propagation, etc.

Fundamental to a conservation policy is the need for an efficient data recording system. Plants are dynamic living entities that require constant monitoring and with endangered plants there is the additional need to record cultural practices and ensure propagation and.distribution. Data recording also increases management opportunities for analysing collections and will hopefully be the future basis for co-ordination between gardens.

ACHIEVING AIMS

Various possibilities have already been suggested. Again looking at Kew it can be seen that in the previous three decades the living collections were curated with passive aims. The emphasis now is on a more active policy endeavouring to make the maximum use of the massive multi-disciplinary resources that are available within the gardens.

Botanic gardens are generally well equipped but often poorly organized to meet the challenge presented by plant conservation. In many instances there is a need for a new organizational infra-

structure and conservation-orientated staff training. Given the
will, however, the massive inertia that has always been inherent
in botanic gardens can be overcome while there is still time to
grasp the grains that are the future harvest for our world.

FUTURE: INTEGRATED INTERNATIONAL POLICIES

R. L. Shaw

Royal Botanic Garden

Edinburgh, Scotland

INTRODUCTION

The more I think about this subject the more incompetent I feel. For a number of reasons a Curator's post these days leaves less and less time for horticulture in association with botany and becomes embroiled in general management matters, many of them wrongly placed and others quite trivial that, in earlier days, required or received little or no attention. Now with a more liberal, even a permissive industrial atmosphere what was quite constant and firm has become as insecure as shifting sand. It takes so long these days to negotiate settlements of many kinds and some settlements are short-lived. Consequently I have insufficient opportunity to be with plants or to influence and encourage colleagues. I find little time to keep abreast of horticultural literature but, very recently, I re-read an article by Professor Heslop-Harrison in the 1971 Journal of the Kew Guild. The title of the article is "The Prospect Ahead" and in it is a precise account of what Kew is doing and what it hopes to do in the future. It is an exciting and encouraging statement and, I believe, a model upon which other botanic gardens might base their activities - if not entirely, then in part. Conservation is not overlooked and nothing would be more appropriate for my contribution to this Conference if I were to read this article to you. Instead I ask you to read it at your leisure.

This leaves me with my predicament and you with the knowledge that at least one member exposed to the limelight of this Conference is not all he should be and that the botanic garden background he emerges from is not all it should be either.

The previous speakers have referred obliquely to inherent
problems. For instance Mr Bruinsma said that our present botanic
garden structures can be blamed for the many problems that beset
botanic gardens today. Mr Simmons referred to "massive inertia"
and botanic gardens often poorly organised. Professor Heslop-
Harrison did not take long to recognise the existence of
parochialism between departments within one botanic garden. I am
not alone then in thinking there are flaws and cracks in the
foundation components upon which we hope to build an international
bridge. If I have anything useful to tell you for the future it
is this: we must put our own houses in order before, or during,
the move towards integrated international policies. Perhaps the
very size of the concept, the tremendous challenge and demands it
will present may be sufficient to catch the imagination and, like
the influence of a magnet, draw our random resources into a
resolute pattern. But the less we leave to chance the better.

COMMUNICATION

At Kew, its Scientific Policy Committee was the magnet and
under its influence there developed a formal communications-network
to end parochialism. Integrated international policy will depend
on a formal communication network and as a start it is essential
that the following information is available.

1. Detailed accounts of the contents of living collections in
   botanic gardens willing to take part in international
   policies.

2. The aims of domestic policies of individual gardens – aims
   that are formulated in the light of prevailing local,
   financial, social and educational circumstances that are
   most likely to remain unchanged.

3. Prompt notification of likely national and international
   expeditions.

4. The dissemination of information on cultural and other
   techniques affecting critical areas of work in botanic
   gardens, eg. record systems, propagation, climate control,
   growing techniques, etc.

Items 1 and 3 should present no insuperable difficulties and
they require immediate attention. Item 4 can only be dealt with
over an indefinite period of time but within 4 there is an element
of urgency to ensure that some system exists to record for posterity
findings and observations the most significant of which should be
publicised. Much more will be said during the course of the

Conference on record systems but I wonder how many of us give sufficient thought to Item 2, to the demands of domestic policies and how they can reduce the contribution of any individual garden to conservation.

INDIVIDUAL POLICY AND ORGANISATION

In South Africa and Australia there is activity towards conservation and we hear of UNESCO's plans for biosphere reserves but in other countries there are botanic gardens with tremendous conservation potential untapped. If this potential were realised a great deal of pressure could be removed from the resources of many European gardens.

I am speaking of countries in the warmer regions of the world and the West Indies comes to mind. In that part of the world, however, the scope of some botanic gardens is severely curtailed for political and economic reasons, and until there is a significant change in the prosperity of these countries, or some form of benevolent external influence can be brought to bear in these regions, opportunities will be missed. For the moment we cannot expect any worthwhile response and we should not be unduly critical of the fact unless help can be offered - the sort of help, I am assuming, that Dominica will receive, or is receiving, via UNESCO.

Coming much nearer home it should not be surprising to learn that while some European botanic gardens will agree entirely with the aims of this Conference they will not be able to contribute very much and again we must try to understand why.

Botanic gardens in Europe come under the control of a number of authorities (some are privately owned) and botanic gardens perform many roles - some conflicting with others. The roles are so varied that the single term "botanic garden" is confusing; at least it is confusing to botanic garden purists, like myself, who believe or believed that any botanic garden should be an institution, no matter how big or small, that gives priority to scientific work, and that anything else attempted has a close link with, or is a by-product of, scientific work. Now, I am not so sure, for there are a number of ways of approaching this subject. Or, I am not so sure that all the activities I am referring to are rightly placed under the single heading "botanic garden".

In Hamburg Botanic Garden for instance it has amongst other things a particular commitment to provide plant material for teaching purposes. In Kew there is a department that until quite recently was entirely devoted to ornamental horticulture. In the Palmengarten in Frankfurt, and justifiably, a bandstand, a fine restaurant and ballroom, play areas and a boating lake are significant features. In Edinburgh a demonstration garden with a herbaceous border jointly display common ornamental plants and in the Schloss Wilhelma in Stuttgart-Bad Canstatt the emphasis is on fauna in a most remarkable way.

Recently in Edinburgh a survey of the public's reactions to botanic garden facilities was carried out and one of the most popular roles seemed to be the provision of peace and beauty. There is even another role in Edinburgh and Kew where in total 100 students leave the gardens as qualified horticulturists every three years.

It is common knowledge in Britain that student training could be done elsewhere (in fact there is considerable competition for students amongst the centres of horticultural training) and again in Britain our local authorities via parks departments cater for recreation and ornamental facilities. There is even some provision for plant material for teaching purposes, while zoological gardens are usually separate from botanic gardens. In short many of the tasks performed by botanic gardens could readily be done elsewhere. But where else are there facilities to compare with botanic garden facilities for certain aspects of conservation work? Are not botanic gardens expected to do too much or have botanic gardens taken on some areas of work because they could think of nothing better to do? Have I not heard on more than one occasion "forget the scientific aspect of botanic garden work - the scientists are not interested in the garden".

I am not suggesting some immediate radical change where all non-scientific work is abandoned but I think it is important that we examine our objectives and place them in some order of priority. In some cases the local restraints I mentioned earlier will force some botanic gardens to place a very high priority on certain duties to the extent that they cannot take much part in conservation.

From what I have read recently I understand that the biosphere reserves programmes have been thwarted to some extent by human politics; from what I have said you will realise that there are far-reaching implications and we must attach a great deal of importance on how and when we can influence the authorities that control the activities of botanic gardens and other major public and private institutions in our countries. I am certain botanic gardens cannot by any means deal with conservation on their own and I shall return to this theme as I develop Mr Simmons' term "Conservation Agency".

Let us assume now that we have given priority to our work, that conservation has a place in botanic gardens, and that we are left with a clear field for botany and a clear view of the plants we grow as scientific collections. What policy governs their

constitutions? Is it one of specialised collections that as far
as possible do not overlap collections elsewhere, or is it a policy
of representation in the widest sense? Or is there a rather vague
policy that in practice is interpreted in a number of ways by
individual members of staff? The latter is not uncommon and I
fear that if the directors of botanic gardens in this category do
not search for and formulate clear cut long term policies and do
little to insist that policy is adhered to then nothing useful can
be achieved. Stringent control is necessary and must be applied
equally to horticultural and scientific staff. There still remains
a gap between these two groups that must be closed. Obviously they
follow two distinct disciplines but even so they should come under
a similar form of control to ensure efficiency, equal opportunity
and compliance with policy. At Kew integration of this kind has
made considerable headway and I was aware of good liaison between
both groups in some German botanic gardens.

REPLICATION

I have had the good fortune to see botanic gardens in a
number of countries and two factors were unmistakable.

1. Established botanic gardens are crowded with plants.

2. Many gardens grow a similar range of plants.

There are only two sensible arguments for this repetition
(1) it would be risky to attempt to grow certain groups of plants
in one place only   (2) the public have a right to see a good cross
section of the world's flora in their local botanic garden.
Argument (1) can be qualified so that repetition is reduced to a
reasonable level and only difficult plants should be widely shared
to ensure their survival. Argument (2) is overstated and many a
collection defended in this way is far too varied and I suspect it
is added to over the years at the whim of an individual who likes
to collect plants as some collectors acquire stamps or matchboxes.
It was Thomas Mann's character 'Felix Krull' who many years ago
said "That is how it is in museums; they offer too much; the
quiet contemplation of one or a few of the objects from their
store would certainly be more profitable for mind and soul". What
he said then is true today for some of our collections offer too
much. They can profitably be reduced.

Only in one or two of the newest botanic gardens in Germany
did I see any free space in plant houses. How then are we going
to find proper accommodation for the 20,000 taxa now threatened in
their natural habitats? In view of financial restrictions and the
energy crisis it is unlikely that we shall see more accommodation
for tender plants. The only solution in that category then is to

make room by reducing the existing plant population in botanic gardens. My immediate solution is to limit the amount of replication that exists. In the same breath there is obviously a very good argument for specialist collections or even for the zoning of taxa so that particular groups of plants can be accommodated in one or only a few places. A compromise is for each garden to develop a number of specialist comprehensive collections and then permit some representation of the world's flora in whatever space remains.

I hesitate to put a figure on the space gained if we could reduce replication but a conservative estimate might be 5%. Kew has usefully put figures at our disposal and on the basis of their 25,000 taxa holding, room could be found for 1,250 additions – about 6% of the endangered species estimated at about 20,000.

The figures I have used so far are not accurate but how accurate can we be with botanic garden figures? Let me use Kew's figures again. Mr Simmons mentioned the taxa held at Kew in 1934 as 25,000. In the Kew Guide of 1961 the figure is still 25,000 and later from the Journal of the Kew Guild in 1971 the figure 20,000-25,000 is stated. When I was at Kew the intake of accessions ranged from 2,000 to 5,000 a year so with a mean input of accessions of 2,000 per year between 1961 and 1971 it might be expected to find that Kew's holdings had increased by 20,000 accessions to 45,000. What happened to those accessions? I know some are dead on arrival (poor seed, etc.), others fail to establish, and another group is used to replace ephemeral species but there must be a gain somewhere, or is there? What happens to accessions in Edinburgh where there is a regular influx each year and where incidentally the new planthouses contain <u>fewer</u> taxa than the houses they replaced?

In both cases Edinburgh and Kew were as sated with plants in 1961 and 1971 as they are now. There is only one answer. For every additional plant that arrives one is lost or discarded. I have used these figures to show that we really do not know much about the losses end of our records systems, but if losses equal gains, what price conservation in botanic gardens?

And while on the topic of facts and figures, how many of the plants that we grow in botanic gardens are correctly named? John Simmons mentioned the <u>Lebronnecia</u> that turned out to be <u>Gossypium</u> and I can recall giving <u>Picea jezoensis</u> particular attention because of its rarity in Britain only to find out much later, when it was possible to get an accurate identification, that it was a <u>Picea sitchensis</u> hybrid. Are there many more plants upon which we waste our resources, thinking we are nurturing valuable material? In Kew and Edinburgh? Yes! In Germany? Yes! In the West Indies? Yes!

How can we possibly expect general purpose botanists and horticulturalists to be so knowledgeable that they can detect critical taxonomic errors in the confusing diversity of general collections? Surely there would be a much better chance of detecting replication and errors of identification if we took more kindly to specialist collections, like some of the excellent collections I saw in Germany. Surely, too, where specialists can concentrate on a limited field, techniques for successful culture are developed more certainly and quickly. I am making these remarks so that we can find room for plants that may otherwise be lost to the world but I am also making them to help conserve those that are <u>already</u> in cultivation.

ECONOMICS

Economics and the replication of material or any other resources are closely linked but Mr Simmons mentioned another aspect of economics and conservation in the form of <u>Fagus sylvatica</u>, a long-lived and amenable species that once established costs little to maintain and produces viable seed off and on for about two centuries. There are in European botanic gardens many trees and vigorous shrubs that could safely be grown outside of botanic gardens and once removed could leave room for species whose cultural requirements and maintenance are more exacting. National Trust gardens in Britain, motorways, parks departments, factories and schools, and other institutions that find it necessary or desirable to improve amenity through the use of plants should be made aware that it would be much more useful and no less attractive to include in plantations species that have been recommended, even supplied, by botanic gardens. With some imagination and with little extra cost a scheme could be devised in which botanic gardens made available properly documentated propagation material of ornamental plants collected from the wild. Multiplication would be the responsibility of the nurseryman, and the clients would bear all the planting and maintenance costs. In return for the original stock and guidance and advice to nurseryman and client, botanic gardens could expect access to all plantations as scientific reserves.

Consider the motorways alone. In Britain between Bristol and Perth and Glasgow and London there are miles and miles of verges rarely disturbed by human beings where countless numbers of individual trees or <u>populations of one species</u> could find ample space. This land contains many soil types, and variations in altitude and climate. This is how I visualise botanic gardens acting as "Conservation Agencies" so that plants can be farmed out to be used as specimen trees, woodland, shelter belts or screens in roads, open spaces, new estates and shopping precincts. Why not? I believe the residents in these areas could be encouraged

to take a sense of civic pride in the fact that their district is
playing a part in conservation.

Botanic gardens will have the knowledge of where the plants
can best be grown and the clients' requirements must receive proper
consideration. We know for instance that <u>Camellia</u>, <u>Forsythia</u> and
<u>Pyracantha</u> grow exceptionally well in London and in their season
are very ornamental. Parks departments and National Trust gardens
might be called upon to accept and maintain, on a permanent basis,
cultivars of genera that befit particular parks or gardens and at
some point specialist floral societies could be involved in genera
like <u>Rosa</u>, <u>Begonia</u>, <u>Pelargonium</u>, etc.

Botanic garden based "Conservation Agencies" could, as time
and conditions permit, extend their activities abroad but initially
great strides could be made in accordance with international policy
at national or regional level. In this way it is possible to
satisfy a number of requirements at one time and this means conservation in the broadest sense.

FOREIGN TRAVEL, PUBLICITY, STAFF

Under the heading of travel, publicity and staff I bring together communication, policy and economics. Obviously travel will
be necessary to acquire plant material, to meet members of other
botanic gardens and to advise botanic gardens requesting assistance.
Publicity as an information service and recruiting and retaining
valuable botanic garden staff will all cost money.

There is a considerable difference between publicity as an
information service and publicity as advertisement. We are all
aware of the tremendous demands commercial advertising makes on
our resources. Advertising companies employ intelligent imagina-
tive people, expensive materials and equipment while the results
of their work, quite apart from disfiguring towns and countryside
alike, prompt us to eat more, wear more, travel more, burn more,
use more, waste more until world resources as we well know become
depleted. Our efforts towards conservation seem puny by comparison
with such extravagance. Cannot the enormous funds used on
advertising be directed, to some extent, towards conservation? At
least might tobacco companies and the rest come to the aid of
conservation as they do to sport?

On holiday this year in the North East of Scotland I came on
<u>Thalictrum alpinum</u> which if not rare is uncommon and where it grew
holidaymakers and golfers trampled it under foot. None of these
visitors to the area was aware of the plant, or aware that they
might be responsible for destroying the habitat of an uncommon
species. Although much has been done to make the public aware of

our heritage, the public remains too ignorant. So there is still
much to do to inform the public who in turn may sway politicians
and moguls alike. Funds are needed for this work and money too is
needed to attract the right sort of people into botanic gardens.
It is no good pursuing advanced and sophisticated ideas if at the
same time other essentials are overlooked. What use is the finest
computerised record system if garden labels are wrongly placed, or
if record numbers are copied incorrectly, or the plants referred
to in record systems are mistreated or neglected or fail to exist?

Mr Chairman, ladies and gentlemen, in conclusion I wish to
quote UNESCO's Michel Batisse who has said in connection with
international conservation programmes, "What is irrational is that
most countries do not understand that their resources are limited",
and, "We are working for the rational utilisation of resources".
In a simple and small way I have, so far as botanic gardens are
concerned, tried to echo these words.

DISCUSSION SECTION II EXISTING COLLECTIONS

Professor Cook (Zurich) stated that the provision of peace and beauty in botanic gardens should not be in conflict with their other functions, and added that small quantities of educative material for the public in a beautiful garden is preferable to a large quantity of education in an ugly garden.

Dr Irwin (New York) agreed with Professor Cook and underlined the importance of a garden as a refuge, especially in high stress areas like New York. He estimated that 90 per cent of the visitors came to the New York Botanic Garden because of its beauty.

Professor Heywood (Reading) said we should not under-estimate the inertia of the public and particularly the scientific community in recognising and understanding the important functions of botanic gardens. If we cannot succeed in influencing trained scientists we will have little success with the general public.

Miss Angel (Kew) pointed out that if a garden was not peaceful and beautiful the public would not come in anyway.

Professor Raven (Missouri) agreed with Professors Heywood and Cook but said that if botanic gardens had not been created for beauty initially they would not have outlived the 19th century. He stressed the importance of education and the role of botanic gardens in conservation but felt that they could not afford the intellectual arrogance of ignoring beauty.

Mr Shaw (Edinburgh) said the importance of beauty was undeniable but felt that it should be a by-product of the scientific work.

Professor Hawkes (Birmingham) compared botanic gardens with museums. The latter have much smaller collections on display illustrating themes by judicious use of material and botanic gardens can usefully copy this technique. Small university gardens such as that at Birmingham have teaching, research and conservation roles. Plantings can be made to tell a story or teach a lesson such as the evolutionary sequence from wild species to modern cultivars in the rose. This can be done in such a way as to keep the necessary peace and tranquillity of the place undisturbed.

Professor Esser (Bochum) agreed that all botanic gardens must
be involved in education and research. More effort should be put
into publicity and botanic gardens can learn from the advertising
world in this respect. They must not educate but must sell
information. Research should not be neglected but the gardens
must be advertised.

Dr Hall (Cape Town) pointed out that Kirstenbosch Botanic Garden
with its emphasis on conservation of native plants in habitat, is
situated on a major pathway to Table Mountain and thus encourages
appreciation of wild areas. Such well-known areas are useful for
convincing politicians of the importance of conservation.

He added that whereas studies of the background of botanic
gardens are so thorough, studies of their aims are very thin, and
asked Dr Budowski whether the IUCN had mapped out a broad brief for
plant conservation on both a world and a regional basis and whether
it was possible to specify critical areas for reserves.

Dr Budowski (IUCN Switzerland) replied that the long term
strategy will be discussed by Mr Lucas (Kew) in his paper, and drew
attention to the IUCN booklet 'Biotic Provinces of the World' which
maps out these areas and analyses how much is protected in each;
for example how little tropical rain forest is protected. He high-
lighted what may be a slight confusion of the role of a botanic
garden in conservation. Should Kew serve as an intermediary in
distributing plants in other areas or should it conserve flora
in situ? If botanic gardens engaged in conservation work, he said
they would receive support from the international programme.

Sir Otto Frankel (Canberra) expressed his interest in the
genetic aspects of conservation. He said that before the public
could accept conservation they would ask: (a) what to preserve,
(b) how to preserve it, and (c) what should be done once it was
preserved? As a geneticist he felt doubtful how far these questions
could be answered. Concerning the plants themselves no continuity
was assured in a botanic garden; even less than in a nature reserve,
it being so easy to let a few specimens lapse. Their ultimate
purpose is for release but it is far easier to release animals,
with built-in homeostasis, than edaphically controlled organisms
like plants, which have to be adapted exactly to their site.

Professor Ehrendorfer (Vienna) said it would be very useful to
prepare a leaflet in which current activities and ideas could be
exchanged between botanic gardens. News of conservation efforts,
education projects, publications and forthcoming expeditions could
be included, the proceedings of this Conference being a start.

Dr Corona (Mexico City) commenting on Mr Bruinsma's paper,
pointed out that the Aztec gardens in Mexico in the 12th and 13th

centuries had played an important role in teaching medicine and had
been well organised.

Techniques of Cultivation

MESEMBRYANTHEMUMS AND THE PROBLEMS OF THEIR CULTIVATION

Hans-Helmut Poppendieck

Hamburg University

Federal German Republic

The collection of the Mesembryanthemaceae is the largest specialized collection of the Hamburg Botanic Garden. It has been set up for the research needs of a group of systematists at the Institute of General Botany, which forms one administrative unit with the garden. The team consists of Professor Ihlenfeldt, Dr Hartmann and myself together with a technical assistant and several undergraduate students, and we are concentrating chiefly on the morphological and taxonomic problems in this interesting group of succulent plants.

The Mesembryanthemaceae as a family was created by splitting up the old Linnean genus Mesembryanthemum into a considerable number of smaller genera. This was begun by N.E. Brown after his retirement from Kew and continued by G. Schwantes in Kiel in Germany and by Mrs Bolus in Cape Town. Originally a subgroup of the Aizoaceae, the Mesembryanthemaceae is now generally considered to be a family in its own right. Nearly 2,500 species have been described so far, the majority of the 1,445 by the late Mrs Bolus within a period of thirty years. Though being one of the larger families of the Angiosperms, the Mesembryanthemaceae (for convenience sake usually abbreviated to Mesems) have a very restricted distribution; all but perhaps a dozen species are confined to South-West and South Africa.

The Mesems are rather uniform in their basic morphological features – decussate leaves, dichasia, showy polyandric flowers, mostly with a pentamerous capsule, etc. – but they show an extraordinary variability as regards to life forms, which range from small shrublets and articulated stem succulents to cluster forming or dwarf leaf succulents as well as to annual herbs. They are

cultivated either because of their interesting habit - like the
so-called flowering stones <u>Lithops</u> or <u>Argyroderma</u> - or for their
showy flowers.   Another point of interest is the unique mode of
dissemination, they are rain ballists, utilizing one of the most
complex fruit structures of the plant kingdom.

Generally speaking, the knowledge concerning this group is
poor, and the taxonomy is in a state of utmost confusion.   There
are several reasons for this only a few of which can be mentioned
here: restricted distribution in a largely inaccessible habitat;
a high degree of morphological variability;   the fact that most
species are self-incompatible, so that the early European investi-
gators had no fruits available and either neglected or over-
emphasised the diagnostic value of the character;   the most important
fact was however that many of the students of this group were gifted
amateurs but without sufficient scientific background, so an unduly
large number of ill-founded taxa have been created.   The publications
of Jacobsen stabilized at least some of the major nomenclatural
problems, but the basic taxonomic work has still to be done.

Our aim, therefore, is to attain a solid basis for a revised
classification of the family, the emphasis of our studies lying on
morphology and evolutionary biology.   It is obvious that this cannot
be done with herbarium material alone, let alone the fact that
succulents give usually very poor herbarium specimens.   Collection
trips have been undertaken in 1969, 1971, 1972 and 1974, on which
we made about 2,500 collections of Mesems.   We try to get samples
from every known locality with at least three living plants to be
taken to Hamburg - additional voucher specimens are usually left
at Stellenbosch Botanic Garden - and a lot of further material like
mass gatherings of capsules and dried or bottled plant material is
also obtained.

The collection now covers an area of 200 m^2 (250 square yards)
with approximately 7,000 pots containing one to several specimens.
At the moment it is distributed over three small glasshouses, but
there is a new glasshouse under construction where we will have
300 m^2 available, so there is still some room for expansion.

It is of course impossible to try and have such a collection
for the whole family, and so we concentrate on few groups only,
especially the Western Cape Province genera.   Even the major taxa
of the Mesems have a distinct geographical distribution, and the
following sub-tribes of the <u>Ruschioideae</u> are endemic to this region:
<u>Mitrophyllinae</u>, <u>Dorotheanthinae</u> and <u>Leipoldtiinae</u>.   Also the sub-
family <u>Mesembryanthemoideae</u> has its centre of diversity here, and
it is these four groups we have under study.   They represent about
forty genera with nearly 450 species, that is one-fifth of the
hitherto described ones.   A list of genera held in cultivation can
be obtained on request, but I think I do not need to list them here.

In addition to what you may call a working collection we have representatives of nearly each genus of the family in cultivation, but this is only for having reference material. Our system of information handling for this collection is simple if not to say primitive, and it consists only of the collector's notebook and a card index which is arranged generically. It is sufficient for the needs of the scientists, and a more sophisticated form of data handling does not seem to be necessary as yet.

Mesembryanthemums have been kept in European collections since the eighteenth century. Most of these long introduced species come from the summer rainfall area of the central or eastern Cape, and are not at all problematic for northern European gardeners. The genera from the winter rainfall districts however are more difficult, especially if they show a marked seasonal rhythm of growth. The light is seldom sufficient in winter to secure a good development. If their requirements are not properly met, it is very difficult to keep them alive for a longer time and nearly impossible to get them to flower. Three points have to be considered above all: the growth rhythm, the lighting and the substrate. The substrate has to fulfill the following conditions, it must be light, it must not keep the water for a long period of time as the roots might rot, and it must not be too sterile, but on the other hand overdoses of nutrients are equally disadvantageous. After some experiments we found out that pure yellow sand mixed with some Perlite is the ideal substrate for our purposes, and we have been using it now for over five years and have not had any trouble. Using a more sterile substrate like sand from sea-shores or Leca-Ton (burnt porous clay) proved to be less successful because it is then difficult to give the right proportions of nutrients. The Perlite serves three purposes: it keeps the soil from hardening, it retains the water without saturating the soil and it serves as a long-term fertilizer. Additional nutrients are only necessary with very fast growing annuals like Aptenia. We like to have the plants in the same pot as long as possible, for every repotting bears the risk of confusing labels, but as the surface of the soil becomes incrusted rather quickly through the development of blue-green algae, we have to change the substrate every second or third year.

As I said before, nearly all our plants come from the winter rainfall area and have the same type of photoperiodic behaviour. They grow and flower during winter, from about May to August in South Africa and from October to January in Germany. All of them need a resting period of at least eight months, and watering them for more than half a year is very damaging as they lose their vigour and then become susceptible to pests. It is not necessary to alter the length of the days by shading or giving extra light after dark. We need however additional light during the day, as the intensity is not sufficient.

For this we have lamps with two tubes which give a light of 4,000 lux and need 130 w. They are controlled by a photo-cell and are switched off when the sunlight is sufficient. The distance between the tubes and the surface of the soil is about 30 cm. (12 inches). Of course they are heating the glasshouse as well, especially on the many warm but cloudy days we have in Hamburg. Last week for instance we had temperatures of nearly 60°C (140°F) in the late afternoon and still 18°C (64°F) during the night which seems to be too much even for species from semi-arid conditions. A lot of the shrubby species shed their leaves. The ventilation in our small houses is unsatisfactory, and we hope to be better off with the new ones. This extra light is very expensive as it is necessary to have one lamp for each square meter to have any effect. I will come to that later.

There are some other problems of cultivation which might be of some interest; raising plants from seeds is easy, except for the annuals, and there are practically no losses if it is done properly, and you can even do it with seeds that have been stored for more than forty years. We have them germinate under semi-sterile conditions on blotting paper, and one to two weeks after germination they can be transferred into small plastic pots filled with the usual kind of substrate. During the first year it is advisable to keep them moderately moist. We place them into a separate box and spray them three to five times a day and water from below about every third week. The best time for this is in September or October, at the beginning of the growth period.

The best time for collecting in the wild is shortly before the beginning of the growing period, about February/March. The plants are still in the resting state and survive the transport - which can take up to six weeks - without any trouble. Another advantage is that the fruits are already ripe but the seeds have not been disseminated yet. In 1969 we collected in August, and we had considerable trouble with the plants, and also the yield of seeds was quite unsatisfactory.

When the plants arrive in Hamburg they are immediately potted and watered. The flowers have been initiated during the summer and in the first year we usually get a flowering which can never be completely matched in the following years. The plants have then to be adjusted to the switch of the photoperiod, and we have to give an extended growing and resting period. After eighteen months they are in the right rhythm again.

We have to go to all this trouble because we want the plants to look exactly as in the wild, but this is of course not always possible. The succession and number of leaves as well as the internodal length can be very elastic under modified conditions, but in the rather uniform environment in the field they are usually an important taxonomic character. In the long run cultivation can change the habit of woody perennials rather drastically, as can be shown with two specimens of <u>Mitrophyllum clivorum</u>, one from the field and the other from exactly the same locality, but grown in Stellenbosch for over forty years. Another point: in the wild, growth is usually restricted by the available resources, whereas in cultivation the plants do not have to compete with others for nutrients or water supply – they seem to grow forever. With the dwarf succulents however we are getting very close to the picture in the wild.

A collection like this one is of course more expensive to maintain than usual, the largest costs on the bill being the staff and the extra lighting. Apart from the scientific staff who shall not be considered we have one technical assistant for work such as gaining seeds from the pods, raising plants from seeds, for photography and keeping records. We also share one gardener with the other people using the experimental houses, but as our plants are easy to look after, one can take only one salary of DM23,000 into account. The price for one lamp is about DM160, and we had to invest DM32,000 for the whole. In a year that means about DM3,000 if you consider damage and repair work and so on. For electricity we have another DM7,000 to pay, so that the additional costs amount to nearly DM33,000 per annum.

Finally I want to turn to the question of whether a collection like this one can assume a role in the conservation of wild species, provided that it is agreed that the Mesems are indeed worthwhile to be kept. There are several points to be considered. Of course it will not be necessary to keep a collection of this size for conservation purposes, and the space for those genera which have been revised will certainly have to be diminished. We are thinking about giving away complete sets of species to those who are interested. But still in the long run it is probably wiser to keep tropical or subtropical plants in a reserve or botanical garden in their original country, or in a place with a similar climate, than in a northern temperate climate under glass at a high capital cost. This expense is only justified for a short period of intensive research but certainly not for the decades that have to be taken into account for conservational purposes. Luckily for our plants at least there are some promising developments in South Africa, especially the recently established Springbok Reserve which after some years with initial problems is now administered under the auspices of the University of Pretoria.

This leads to the problem of a long term policy, and it is questionable if even the European gardens are able to guarantee long term stability. It must be recalled that specialized collections – some of them very valuable – have been accumulated many times, but they have usually been lost after the gardener or

scientist in charge has died or left the place. This occurs
commonly in smaller botanic gardens particularly if, like our
Garden, they are firmly linked to a University. Such gardens have
to serve only restricted needs and are mostly not able to set up
an independent policy. When the professor retires his successor
may be interested in another group of plants, grasses, instead of
orchids for example, a drastic change in the composition of the
living collections is necessary. But even if one is willing to
keep a large and valuable collection instead of deliberately
letting it go, the prospects are still very grim. Such 'unwanted'
plants are frequently moved around when their space is needed, no-
body really cares for them, the labels get confused in a surpris-
ingly short time and the plants die or at best, are given away. I
do not think that Professor Heslop-Harrison's vision of the
"geriatric ward with only terminal cases" - "for the plants that
have nothing left to offer" will ever come into existence, because
so many large and valuable collections which indeed had something
to offer have disappeared - I am thinking only of the begonias of
Irmscher which we had in Hamburg!

     This shows that the motivation and a sense of purpose is also
necessary for the European gardens, and it can only be hoped that
the motivation will be increased if there is some international
confirmation that it is really worthwhile to keep the plants that
are now in cultivation. It might also help the gardens if they
have to appeal to their sponsors for more money - or perhaps more
realistically in the face of decreasing public spending - to be
spared from severe reductions. It is getting increasingly difficult to advocate the needs and tasks of our institutions with
politicians who think of botanic gardens only as glorified public
parks, nice to have in prosperity but dispensable in times of
financial crisis.

THE ARCTIC GLASSHOUSE

T. Böcher

Copenhagen University

Denmark

The Arctic Glasshouse in the Copenhagen Botanical Garden was built in 1959. For 15 years it has been used for cultivation and experimental work with more than 100 different species, mainly from Greenland. A short paper on the history of the house, some technical details and the prefloration periods of the species grown in the house appeared last year in the volume of "Botanisk Tidsskrift"published on the occasion of the Centenary of the Garden. A more elaborate survey of the technical arrangement has been translated into English and is available on application to Mr H.P. Hjerting of the Copenhagen Botanical Garden.

Let me first emphasize that the house is not an arctic phytotron comparable to phytotrons constructed for physiological experiments, but it provides us with ample possibilities for observing and working experimentally with arctic plants. The house is rather small, 90 sq.m. with a bench area of 40 sq.m. It is divided into four sections, each with separate temperature controls. Each section has a basement where the ventilation equipment is placed and in the winter the plants are stored here in constant frost and darkness. The upper greenhouse sections are covered with two layers of glazing. The space between the two layers is ventilated by cool air from the basement. The air also passes from the basement through floor grilles into the greenhouse. It is returned through a return air-duct and recirculated. The photoperiod is extended by using eight 400 Watt mercury vapour lamps.

In the winter, from late October to the first days of May, the plants are covered with plastic in order to maintain the humidity. The temperature is about -6°C. The floor grilles are covered with shutters so that the conditions for the plants resemble those under a thick snow cover.

Mainly we have been using two Summer climates, a cold one in the first room where the temperature does not exceed 9°C during the day, and a warmer one allowing day temperatures to rise to, but not exceed, 16°C.

The plants are watered evenly, but certain species are maintained in waterlogged soil conditions, and others are placed in plastic tubs filled with Greenland heath soil overlaying coarse gravel. Among the first are species like Phippsia algida and Cardamine nymanii, among·the latter Empetrum hermaphroditum, Cassiope tetragona and Oxycoccus microphyllus.

Several plant species are grown from stations in different parts of the arctic region. They exhibit the same type of ecotypical variation that is well known for temperate species. Thus, there are conspicuous differences in height and vigour in species like Ranunculus sulphureus, R. pygmaeus, Phippsia algida and Alopecurus alpinus. The two latter are astonishing. In nature they both occur in wet soils which may be manured or unmanured. Those growing in manured soils are much more vigorous, which seems very understandable, but the curious fact is that the taller growth of the plants from manured places is maintained in both species when cultivated under uniform conditions in the arctic house. They seem to have developed special races adapted to soils manured by sledge dogs and to real arctic, unmanured soils along melting snow and ice.

It has been of particular interest to compare arctic and alpine populations of the same species. At an earlier occasion I have mentioned the divergencies found in Ranunculus glacialis, among which the differences in the branching of the inflorescence, the breadth of the leaf segments, and the vigour of the plants are conspicuous. The North Atlantic populations appear more uniform than those from the Alps, they are smaller, less inclined to flower, have very rarely more than one flower, and have narrower and more petiolate segments. A similar behaviour was found in Arabis alpina which is more uniform in the North Atlantic area compared with southern mountains. Even Juncus trifidus show this uniformity of its North Atlantic stocks. However, there are many important exceptions; a species like Viscaria (or Lychnis) alpina is abundantly different if populations from Iceland are compared with those from Greenland and Canada. Those from Iceland resemble the edaphically specialized races from Scandinavian serpentine soils or the Alvar race from Øland in the Baltic Sea. The alpine or Scandinavian races, on the other hand, are clearly closely related to the North American-Greenland populations, although the latter are considerably coarser with larger flowers and broader leaves.

In one case the difference between the populations in the Alps and the Scandinavian-Greenland populations was estimated to be a difference between two species. The species pair is <u>Braya</u> <u>alpina</u> which occur in the Alps and <u>Braya linearis</u> in Scandinavia-Greenland. Crossings of these species, when undertaken in the Arctic Greenhouse, showed that the species are interfertile. In F_2 the segregation is very large and covers all kinds of intermediate biotypes, as well as a good many which almost match the parent species; some few F_2 plants, however, are poor or dwarfish and may represent less efficient gene combinations.

   <u>Braya</u> species are difficult to grow because they often die after luxuriant fruit-setting. <u>Braya purpurascens</u>, which is a high arctic species, is unable to stand Danish summer conditions even when grown in pots and watered abundantly, but it is happy enough in the Arctic House. It is just one example among many of the arctic species which simply cannot tolerate Danish outdoor conditions; other examples are hygrophytes like <u>Ranunculus sulphureus</u> and <u>R. nivalis</u>, <u>Saxifraga cernua</u> and <u>S. rivularis</u>. Among xerophytic species <u>Lesquerella arctica</u> and many entities within the <u>Papaver radicatum</u> complex usually also die. However, it is a striking fact that a great number of arctic xerophytes are able to stand Danish summers when grown in pots and watered regularly. Among the best examples are <u>Draba subcapitata</u>, <u>Draba nivalis</u>, <u>Draba glabella</u>, <u>Draba cinerea</u>, <u>Draba aurea</u>, <u>Potentilla hookeriana</u> and <u>Potentilla nivea</u>, <u>Erigeron compositus</u>, <u>Artemisia borealis</u> and <u>Papaver radicatum</u> ssp. <u>subglobosum</u> from northern Norway. <u>Melandrium (Gastrolychnis) triflorum</u>, which in nature grows on dry soils, is able to grow out of doors but is not happy there, whereas the mesophytic <u>Melandrium affine</u> usually dies. When the latter is grown in the Arctic House, we have noticed large genetic differences between plants from Wrangel Island in N.E. Asia and those from Greenland. The arctic Asian strain is smaller, flowers more easily, and has more inflated calyces.

   Among the species studied in the Arctic House there are good reasons to mention <u>Draba sibirica</u>. From two British botanists, who were visiting Hurry Inlet near Scoresbysund, I received living material of this very rare plant. The plants were grown together with <u>Draba sibirica</u> from Greifswald and Moscow. While those from central east Europe have ascending shoots, large leaves and flower twice, in the Spring and in August, those from Hurry Inlet keep adpressed to the soil, have small, dense and more fleshy dark green leaves, and flower only in the Spring. According to herbarium studies similar plants occur at Waigatch, norther Urals, and Novaja Zemlya. It was clear to me that the East Greenland plants belonged to a separate arctic taxon ranging from arctic N.W. Asia to East Greenland; accordingly, it was described as ssp. <u>arctica</u> of <u>Draba sibirica</u>. Without cultivation it could easily have been considered merely an arctic dwarf modification, which is not the case.

The presence of interspecific hybrids between <u>Chamaenerion</u> <u>latifolium</u> and <u>angustifolium</u> were already assumed by M.P. Porsild, but the existence of such a hybird was not shown until quite recently in material from Disko cultivated in the Arctic Glasshouse. This is sterile with 2n = 54, thus a number in between the parents which have 2n = 72 and 36. New investigations show that it occurs in several stations along the coast of West Greenland. The hybrid is probably always formed with <u>C. latifolium</u> as the mother plant because <u>C. angustifolium</u> is sterile under present conditions in Greenland. The hybrid and the parents grow luxuriantly in the Arctic House, but it is a curious fact that Greenland specimens of both species are poor and difficult to grow outdoors in Denmark. <u>C. angustifolium</u> from Greenland attains heights of about 30-40 cm. only, but it produces ripe seeds in pot culture in Denmark. The Greenland plants of <u>C. angustifolium</u> thus belong to a low arctic race which is happy enough in the Arctic House but unhealthy outside.

Another interspecific hybrid is <u>Saxifraga nathorstii</u>. It is endemic to North-East Greenland. It is an alloploid with the yellow flowered <u>Saxifraga aizoides</u> and the purplish flowered <u>Saxifraga oppositifolia</u> as its parents. This was shown many years ago by cytological methods. Now we try to produce this hybrid artificially. So far we have not succeeded in getting the primary hybrid, but we are still hoping to make it, if the few seeds we have harvested after artificial pollination will germinate. It is striking that this hybrid is not found eg. in the Alps, Scandinavia or Canada where the parents are sympatric. Perhaps it arose suddenly by fertilization of unreduced sexual cells produced under the high arctic conditions of N.E. Greenland where it grows. There has been time enough for its establishment, however. According to recent geological data the area has not been covered by ice for the last 40,000 years, and it harbours many rare and isolated plant species, among them <u>Draba sibirica</u>, already mentioned, the arctic <u>Potentilla stipularis</u> and the endemic <u>Braya intermedia</u> which seems to be a triple alloploid species.

The Arctic Glasshouse is clearly of vital importance for all work on the evolution of arctic species. Here we can compare structure and behaviour - life cycles - under nearly controlled conditions and keep some of the rare species growing. Of course, in the vast arctic areas hardly any species is in fear of being eradicated, but some are very rare - perhaps relics - restricted to a few stations far from one another. This is the case with <u>Potentilla ranunculus</u>, a completely glabrous, glaucous species related to <u>P. crantzii</u>. It is now growing in the Arctic House.

However, it may sometimes be even more important or alluring merely to cultivate Greenland populations of some species. Thereby we may be able to disclose secrets and understand the

fate of a species. We can compare the Greenland gene pools with
those from outside and be able to estimate depletion of biotypes,
thus reduction of the gene pool due to severe conditions during
the Ice Ages. As an example I may mention that <u>Dryas octopetala</u>,
which grows in isolation in North-East Greenland is very difficult
to grow outdoors in Denmark when the plants originate from
Greenland, but members of the same population can be maintained
in the Arctic Glasshouse. Thus, the conclusion may be that the
<u>Dryas octopetala</u> from N.E. Greenland is physiologically adapted to
real arctic conditions and not able to grow in the low arctic,
oceanic South-east Greenland.

Of course, our next steps should preferably be ecophysiological; we must work experimentally with the species along lines
similar to those followed by Billings and Mooney in their very
inspiring work on <u>Oxyria digyna</u>. But we must not forget the
detailed structural analysis. Warming and his contemporaries
started working on the structure and flower biology of arctic
species and this should be followed up by new methods. What is
the function of the peculiar shield hairs in <u>Rhododendron lapponicum</u>?
How can we understand the multicellular glands that cover the
dorsal side of the leaves in <u>Cassiope tetragona</u>? Experiments are
needed, and we must be able to walk just downstairs and into the
Arctic Glasshouse to collect material for the experiments or for
histochemical, light microscopical, or electron microscopical
observations.

Although many arctic species are kept in culture in the Arctic
Glasshouse it is clearly too small to fulfil the demands of a
steadily increasing scientific activity on the arctic flora. Any
dream of being able to cultivate just small populations of the
same species from many parts of its circumpolar range must of
course be abandoned. We must be content with taking up a few
species for large scale circumpolar investigations. However, we
can increase the number of cultivated species somewhat in one way:
by keeping the plants outdoors in frames during the summer, bringing them into the Arctic House during the winter and placing them
there in constant frost and darkness. In fact, many species are
easy to grow in this way, except for snow patch species and many
species from flushes, marshes and wet clay. In the future, we may
proceed in this way, but we may further, if the financial
situation improves, be able to establish small experimental gardens
in some of the Greenland towns. My vision is three such stations
at different latitudes: a subarctic, a lowarctic and a higharctic
station, and to try to cultivate not only arctic species in these
gardens but also for example certain temperate species or alpine
species. As a first approach I have in 1974 transplanted Danish
strains of <u>Chamaenerion angustifolium</u> and <u>Campanula rotundifolia</u>
to a very small piece of cultivated soil just outside the Arctic
Station of the University of Copenhagen at Godhavn, Disko Island,

in West Greenland. But for an ageing man like myself such business
is rather like planting trees for future generations.

TECHNICAL DESCRIPTION OF THE ARCTIC GLASSHOUSE (GREENLAND HOUSE) IN THE BOTANICAL GARDEN IN COPENHAGEN

Olaf Olsen

Copenhagen Botanic Garden

Denmark

In the arrangement of the arctic glasshouse the variations in climate from South to North Greenland have been taken into consideration. For practical reasons the glasshouse has been divided into four sections corresponding to four climatic zones. During the four summer months the following maximum temperatures are observed here:

	Day temperature	Night temperature
Section I	6°C	1°C
Section II	8°C	2°C
Section III	10°C	3°C
Section IV	12°-14°C	4°C

The design of the glasshouse is based on the following maximum temperatures.

- 2°C indoors   in winter, dry bulb thermometer

+ 12°C indoors   in summer, dry bulb thermometer

+ 30°C outdoors   in summer, dry bulb thermometer

+ 20°C outdoors   in summer, wet bulb thermometer

Corresponding to the conditions in nature the temperature is stepwise adjusted after winter storage of the plants. The lowest temperature in winter is -7°C.

The glasshouse has a basement storey equipped for winter storage and for accommodation of the evaporator elements, which are connected to the four sections of the glasshouse through floor grilles extending throughout the length of both sides of the glasshouse. In this way air circulation is maintained and the required temperature control ensured.

In each section there are two fluorescent mercury vapour lamps (Philips HPLR 400 Watt.). This is probably slightly too little. The luminous power is 19,000 lumen.

Refrigerant: Dichlorodifluoromethane, CCl_2F_2, Freon 12.

Lubricant: High-viscosity lubricant, since Freon 12 reduces the viscosity of the lubricant applied.

CONSTRUCTION AND INSULATION OF THE GLASSHOUSE

Outer walls and floors are of concrete, thickness of wall 14 cm, in the basement 30 cm. Inside of walls covered with 8 cm foam plastic. On the ground there is a 10 cm layer of slag, 10 cm of mass concrete, 10 cm of insulating material, and a 10 cm finishing layer of concrete.

The glass roof is double glazed and the partitions are of glass in wooden frames.

The orientation of the building is with gables facing north and south to reduce direct solar radiation. The slope of the roof is 37°. Total loss of radiation amounts to 80,000 kcal/h.

Evaporators and Ventilation

In Greenland the relative air humidity is on an average 75%. With a temperature in the room of $-6^{\circ}C$ ($\pm 1^{\circ}C$) air is blown into the refrigerator at $5^{\circ}C$ and with a relative humidity of 70%. The volume of air circulating is 15,400 m^3/h with direct expansion. The finned evaporator is built into a casing to guide the air flow. A fan (two speeds) blows air through the evaporator to provide the required cooling.

The evaporator has ducts for guiding the circulating air and from these it is blown out below the ceiling of the basement, which has grilles through which it enters the glasshouse.

Through the return air-duct in the middle of the glasshouse the warm air is returned to the plant and again cooled.

For the space between the two layers of glazing air is drawn in from the outside through grille and a duct below the basement floor and passed through the refrigerating system. The air for this purpose amounts to about 10% of the air for ventilation and corresponds to half of the volume of the glasshouse per hour. It escapes through a slit in the ridge of the room.

The fans are operated by two-speed motors. At full speed the fan capacity is equal to an air change of 140 times per hour.

The condenser is an Atlas Recold Evaporative Condenser with combined air and water cooling. It consists of four sections, each with its own compressor.

Action of the Automatic Equipment

The glasshouse refrigerating plant is divided into four equally large sections. Refrigerant evaporates in the finned evaporators, and the gas is removed by the compressor leaving room for further evaporation.

In the compressor the gas is compressed and from here transferred to the condenser section of the evaporative condenser. Here the gas is condensed and again returned to the finned evaporator. A special vibration eliminator is placed in the suction and pressure line.

Each compressor is fitted with an oil separator with automatic oil return (sight glass – check valve).

The compressor is provided with a relief device operating in case of overload (interruption of current). Impurities and water are retained in a wet-dry filter.

Magnetic valves close the pipes while defrosting takes place. The supply of fluid to the evaporator elements is controlled by a thermostatic expansion valve fitted with a fluid distributor. The suction line is kept dry by a heat exchanger.

Defrosting

Below each evaporator element an electric heater is fitted which through an automatic clock for defrosting puts the plant out of operation. The setting of the clock determines the duration of the defrosting period.

Control of Air Temperature

The air temperature is adjusted by means of a Honeywell thermostat. The latter is connected to an electrical panel, a step regulator with built-in potentiometer. From here an impulse is transferred to an electrically operated valve situated at the suction side of the evaporator element. In addition the impulse from the thermostat acts on the compressor.

When cooling is required the electrically operated valve opens automatically, and the temperature of the evaporator element is lowered. If the required air temperature is not obtained with a fully opened valve, the automatic control system will act upon the compressor so as to cause its output to increase.

As the temperature of the atmospheric air changes with the seasons a Honeywell control device is provided enabling the adjustment of the system according to the different requirements of summer and winter. It is also automatically adjusted according to the temperature variation from day to night.

Choice of Glazing

Originally a greenish type of glass, "Filtrasol", was used in order to reduce harmful temperature variations. It was found, however, that too much light was absorbed, the arctic plants being subject to etiolation and elongated growth. Next year normal glass was consequently substituted.

Procedure

During winter the plants are covered with two plastic sheets to reduce evaporation to a minimum. At the beginning of May the plants are transferred from the basement to the glasshouse, and in the course of October they are again returned.

Temperature Regimes for the Arctic Glasshouse

		Night and day temperatures			Times of day/night temperature shifts for Sections I-IV	
		Section I		Sections II-IV		
		Night	Day	Night	Day	Night to Day	Day to Night
1st Week		1	1	1	1	–	–
2nd	"	1	3	2	5	4.30	19.30
3rd	"	2	5	4	9	4.00	20.00
4th	"	2	7	5	13	–	–
5th-14th	"	3	9	6	16	3.30	20.30
15th	"	2	8	5	14	4.00	20.00
16th	"	2	7	5	12	4.30	19.30
17th	"	2	6	4	10	–	–
18th	"	1	4	3	8	5.00	19.00
19th	"	1	3	3	6	5.30	18.30
20th	"	1	2	2	3	6.00	18.00
21st-etc.	"	1	1	1	1	–	–

Times at which Lamps are switched On and Off
in the Arctic Glasshouse

	Sections I to IV			
	Morning		Evening	
	On	Off	On	Off
1st Week	3.00	5.30	18.30	21.00
2nd "	2.30	5.15	18.45	21.30
3rd "	2.00	5.00	19.00	22.00
4th "	1.30	4.45	19.15	22.30
5th-14th "	1.00	4.30	19.30	23.00
15th "	1.45	4.45	19.15	22.15
16th "	2.30	5.00	19.00	21.30
17th "	3.15	5.15	18.45	20.45
18th "	4.00	5.30	18.30	20.00
19th "	4.45	5.45	18.15	19.15
20th-etc. "	–	–	–	–

THE PROPAGATION OF HAWAIIAN ENDANGERED SPECIES

Keith R. Woolliams

Waimea Arboretum

Hawaii, U.S.A.

INTRODUCTION

The following summary of the state of the Hawaiian flora, will serve to show the urgent necessity of it's preservation:

Total number of Taxa:	2,734	(Endemic:	2,668)
		(Indigenous:	66)
Endangered	800		
Status unknown	518		
Extinct	273		

By regarding 'status uncertain' as endangered until more is known, we have a total of 1,318 - or almost 50% of the known species either under stress or facing extinction - a sad state indeed.

To complicate matters further, we do not know just how many species are endemic to Hawaii. The above figure represents a summary of present knowledge though Degener suggests there may be as many as 20,000 species!

Reasons for the Present Situation

Inevitably man is largely to blame through his destruction of habitats over a long period of time. While it is true that State and National Parks are providing valuable protection to many areas, the majority of the remaining Hawaiian flora is under considerable

73

stress - mainly due to the depredation of feral goats, pigs and
sheep and the rapid spread of destructive introduced weeds such as
Passiflora mollissima, Rubus penitrans, Rubus ellipticus and
Clidemia hirta. That these can be controlled but are not, due to
political and economic pressures, pinpoints the plight of Hawaii's
flora today. A case in point is a recent report in a local news-
paper that goat hunting on parts of Kauai will be banned during
1975 to "allow populations to increase to satisfactory hunting
levels". Meanwhile the long battle to prevent the introduction of
Axis deer onto the island of Hawaii is far from over.

Not all plants are facing extinction because of man however.
In a flora which has evolved in very specific habitats, it is
inevitable that small changes in, for example, climate can have
marked effects.

In addition, there is little doubt that some species are
simply facing the "end of the evolutionary line" and are naturally
becoming extinct - as species have for millions of years.

During the last 20 years, work on the cultivation of Hawaiian
endemics has been carried out by a few people in Hawaii, though
few records of propagation have been kept. Obata published two
articles on his experiences and plants of several species are in
cultivation at Lyon Arboretum, and the Honolulu Botanic Garden.

AIMS OF CULTIVATION

While there is no doubt that in most cases habitat conservation
is the most successful method of species conservation, in the
absence of such, preservation through cultivation becomes essential.

Our aims in this work are:

1. Preservation through cultivation in botanical collections
   and amassing data on cultivation techniques.

2. Re-introduction into the wild under scientific control.

3. Popularizing the more ornamental species for home and
   landscape use.

Another important aspect of our work is education; through
tours, displays and publicity we hope to spread knowledge of the
importance of preserving the worlds of flora and fauna.

Collection of Material for Cultivation

We have found that it is usually easier to propagate from plants in cultivation than from material brought in from the wild. The reason for this is that proper selection of seeds or vegetative material can be made. When collecting in the wild there is often little choice of vegetative material and fruit and the correct stage for harvesting is not always available. In addition, there is an inevitable 'shock' caused to vegetative material by the delay in getting the material to the propagating area plus introducing it into a different microclimate. Where rare species are concerned, one has to balance the chances of success with such material against the possible effects on the wild population of removing it. This is especially true when seedlings are taken from the wild as results are often unpredictable.

For these reasons it is useful to grow a wide range of species - including commoner species - in order to get knowledge and experience to help cope with problems which might arise with the rarer species.

When an expedition is going to collect living material over a period of time, it is important to devise a method of keeping it fresh until arrival at the nursery. A satisfactory method is to prepare cuttings and plants at base camp; wrapping basal parts in moist sphagnum moss and enclosing them in a tightly tied plastic bag. The aerial portions are left exposed. These are then laid in rows - not more than two deep - in large plastic bags; with the top folded over. Conditions in this plastic 'tent' are controlled daily by opening and closing the bag. When the inside condensation is too great the bag is turned inside out. In this way cuttings and plants can be kept in good condition for up to three weeks. On short trips, it is enough to carry everything in a large plastic bag.

When taking cuttings it is good to get a wide variety of wood; with seeds, unripe fruit of mature size will often ripen if they are placed in a plastic bag (with a few holes in it) and hung up out of direct sun. After ripening, the seeds can be sown in the usual way.

It is important that horticulturists become familiar with a wide range of habitat types by collecting widely themselves. A disadvantage of course is that they will not always be familiar with the species - especially in such a diverse flora as that of Hawaii.

If persons other than horticulturists are collecting, it is essential that information which can help successful cultivation is noted and accompanies the material. Such information as for

example, the percentage of light, soil drainage, water requirements and air movement can sometimes be of immense value. When
working "blind" the chances for success are drastically reduced.
For example - to receive cuttings "collected E. Maui, elev.
2000 ft.", does not really help very much!

Obviously the ideal is for horticulturists and botanists to
work closely together. In this respect, we at Waimea consider
ourselves exceptionally fortunate in having Dr Derral Herbst as
our honorary consultant for Hawaiian flora. Dr Herbst knows the
plants and appreciates the needs of the horticulturists.

Propagation Facilities

At this stage it may be useful to try to describe the propagation facilities at Waimea.

The area is well protected as it is located at the foot of a
steep bluff which bisects the valley; however there is plenty of
air movement.

We believe it is essential to keep the propagation area small,
so that a constant watch can be kept on all plants. To draw
attention to endangered species, a red line is made on all "lead"
labels - ie. the first label in the row. This is a ten-inch
plastic label.

Seed House

This is an insect-proof house with a solid roof of green
corrugated plastic with filtered light and screened sides. Watering is by manually operated mist. The bench is of heavy wire
mesh and one portion has a heating mat with temperature controlled
by thermostat. The floor is of concrete which aids hygiene.

Cutting Beds

Vegetative propagation is mainly done in a long bed built of
concrete blocks on a concrete floor base under 63% shade cloth.
This bed has six divisions of pure coarse sand or cinder and peat,
over a layer of coarse gravel, and with or without heating cables.
Mist is operated by an 'electronic leaf'. Hardening-off after
potting and prevention of transplant shock is provided on benches
with time-clock controlled mist. A solid roof for rain protection,
covered with roll-up shade cloth provides either 10% or 73% shade.
In order to provide rain protection or additional shade, mobile
shelters can be rolled over any bench.

The main benches for growing plants are either under 30%, 55%, 63% or 70% shade cloth to provide a wide range of conditions. All benches are of slotted angle-iron with slats on top; coarse gravel is below. Paths are of concrete.

Set in the roof of the propagation area are adjustable manually operated nozzles which can provide mist for humidity increase or can be used for watering in the summer in long dry periods.

All floors of the propagation area are treated with a sealant and washed weekly.

PROPAGATION TECHNIQUES

Seeds

We usually hot-water treat or file hard seeds and we are just beginning to experiment with gibberelic acid soakings.

Sowing is usually in our standard two parts soil, one-part peat (medium), half-part perlite, quarter-part black cinder; the soil is a mature mixture of chicken manure, soil and shredded leaves and chips, partially sterilized at 160°F.

Where practical, seeds are counted when sown and first germination is recorded. All information is noted on the back of the lead label.

We are beginning trials with sterilized and unsterilized soil in cases where germination is known to be rare, as almost nothing is known of possible mycorrhizal or other relationships. Another unknown is to what extent an after-ripening period is needed before germination can take place.

On several occasions seeds from the same plants in the wild, apparently fully ripe, consistently failed to germinate under any of our treatments. Then for no apparent reason, another harvest would give an excellent germination. Just why this should be is not known, but more work needs to be done on this. Flask culture (ie. absicion of embryo) should give interesting results.

On other occasions, a first sowing yielded no results, but a second sowing a few months later, using surplus seeds from the refrigerator, produced a few seedlings!

Green <u>Pritchardia</u> seeds, providing they are of mature size, will usually germinate. The time needed for germination appears to

be longer than for fully ripe seeds - as may be expected. Green
Pittosporum fruit will often open in a plastic bag and after sowing
in the usual way may germinate.

There is no doubt that we have a great deal to learn about
the physiology of Hawaiian seeds. From the practical point of
view we do not yet have enough data recorded to draw any meaningful
conclusions.

A final word while on this subject: I know of no-one who has
successfully germinated "Pukiawe" - Styphelia tameiameiae - yet
this plant is one of the commonest low to high elevation plants,
and seedlings are abundant in the wild. Possibly the fruit passes
through a bird .... so we are now considering using a mynah bird
as part of our propagation programme!

Cuttings

The cutting-bed is under 63% shade, as full sun is too strong
under our conditions of long hours of clear skies - especially in
summer.

A large number of Hawaiian plants grow in shady conditions and
results indicate that cuttings from such plants root better when
approximately the same amount of shade is given during propagation.
Rooted plants are more tolerant of variation.

Use of rooting hormones containing Thiram fungicide is standard treatment - but trials with and without hormones are often
made. On one or two occasions, hormones seem to have had detri-
mental results, but this has not been positively established and
results are not consistent.

Where cuttings take a long time to root, liquid fertilizer is
watered on every two or three weeks. More trials on the use of
this are planned.

When the species is rare, every piece of available material
is used and many kinds of cuttings are made - usually with
inconsistent results. On balance, however, heel cuttings do best
for most species.

When rooted, the plants are potted into the standard potting
mix of three parts soil (unsterilized), one part cinder, half part
perlite; however, this is varied when necessary.

The pots are placed under clock-controlled mist and occasionally Captan or Tri-copper Sulphate drenches are used as precautions
against damping off and stem rot (Pythium), which are prevalent in
our area.

AFTER CARE

As might be expected, plants from elevations of for example, 3000 ft., do not do too well at or near sea level, and to date no botanical garden in the islands has a high elevation site. However, we have found that if the plants are given say 10%-20% higher shade than they get normally, to reduce leaf temperature, and plenty of aeration is provided, they can be induced to survive for quite a long time.

At Waimea, the arboretum site goes from sea level to 1000 ft. and rainfall from twenty-five inches to over one hundred inches per annum. Although a site at 3000 ft. would be a great advantage, we feel that even at 1000 ft. there is a good possibility of several of the higher elevation plants from wetter areas growing quite well. Those from drier areas may grow at lower elevations if a little extra shade is given.

The problem is that many Hawaiian endemics are unpredictable from the time they arrive at the propagation area until they are flowering and fruiting months later. The main reasons for this are nematodes and fungus diseases - especially Pythium stem rot - which appear to be more prevalent in the lower elevations.

Obviously this has to be taken into consideration, especially with dry-land plants, for example Sesbania species, where continuously damp soils aid the rapid spread of soil fungi. By using sand and gravel mixed with the soil and layers of sand and gravel on top of the soil, plus occasional applications of Captan and nematicides, some degree of control can be obtained. Drip irrigation shows promise also, as the soil surface can be kept relatively dry. This is especially good for strand plants as in the wild the neck of the plant is naturally almost always dry, while the roots never are.

For more moisture loving species, a microclimate which is shady, cool, humid and has mist nozzles to try to imitate the "cloud-drip" found in many habitats, is being constructed. No doubt much experimentation will be required. It is hoped that many of the rain-forest Peperomia species, epiphytic ferns and lobeliads can be grown in such an area.

Hawaiian plants are very susceptible to damage by insecticides and fungicides and will often die more rapidly through their use than without! The same is true to a lesser extent with fertilizers. In the former case, it is necessary to test on a few plants first, while in the latter case the purely organic "soil conditioner" type fertilizers or liquid feeding, produce the best results.

To sum up, perseverance is necessary throughout and large

numbers of plants should be grown where possible, to allow for
sudden losses.

Examples of Successes and Failures

Hibiscus: All the endemic Hibiscus are easy to grow, either
from seeds or cuttings (though H. kahilii is sometimes difficult
to root). As most of them are floriferous and attractive their
future seems assured, though they are not yet widely cultivated.

The seeds respond well to a warm water soak prior to sowing,
while the use of rooting hormones is a definite advantage for
cuttings. The higher elevation species do quite well in the low-
lands and several of the low elevation species, eg. H. rockii,
which is restricted to a single roadside colony on Kauai, grow
better in cultivation and respond well to feeding.

Argyroxiphium species: the "Silverswords". These will not
succeed at low elevations except in very special circumstances and
with a lot of care, especially with watering. As a guide, it
might be said that if Protea can be grown, so can Silverswords.
A very porous soil is needed; little water (though the plants
receive much "cloud drip" in the wild) and an almost continuous
breeze. A. kauense comes from a naturally moister area than
A. sandwicense, the better known species, but is still a high
elevation plant. In the low elevations a cooled house with fan is
the best method of cultivation.

A. virescens, the Maui Greensword, is a high elevation species
from rain-forest habitat and appears to be less particular in its
requirements. Good drainage is essential and 10%-20% shade to
keep off the extreme heat of the sun, is beneficial.

Pritchardia: An interesting project during the last year has
been the mapping of known colonies of Pritchardia, the only
Hawaiian genus of palms. This group is taxonomically very confused
and many of them are highly endangered, for example, P. monroi is
known from only one (possible two) plants on Molokai. Plans are
being made to gather quantities of seed of this species, through
the help of the Department of Fish and Game. As rats eat the seeds
of all Pritchardia netting of the inflorescences may become necessary. Pritchardia kaalae and it's variety minima are only found
as small colonies on Mt. Kaala, Oahu. They are now in cultivation.

Other Pritchardia grow often on inaccessible slopes or tiny
islets with sheer cliffs, often with no landing place for a boat.
While we can guess at their affinities, we cannot be sure and we
hope to raise funds for a helicopter for collecting from some of
these difficult areas.

As mentioned before, Pritchardia are not difficult to grow and green fruit sown direct or allowed to ripen in a plastic bag prior to sowing, usually germinate. Although most of the species are either coastal plants receiving much salt spray, or high elevation plants often from windswept mountain ridges, they do not appear to "need" these habitats and will grow well in the lower elevation in cultivation.

P. remota, from Nihoa Island, is known from two valleys there. In the wild many of the seeds are "sown" by the shearwaters when they burrow in the sand to make their tunnels.

Sesbania tomentosa: This species is of special interest because it is found as several island forms, that on Hawaii Island with prostrate branches and orange-pink flowers, and the prostrate Molokai form with red flowers, being the rarest and in need of immediate protection. The Molokai tree form .... forma arborea, is also extremely rare and endangered.

By hot water treating the seeds prior to sowing and then repeating this over and over again for those which do not germinate, almost 90% germination has been achieved over a period of several months, even with seeds over two years old.

As far as cultivation is concerned, it is very susceptible to Pythium attacks, but with attention will flower and fruit well. If planted in very dry localities or by having young plants constantly ready for planting, it can possibly be kept in cultivation. It is too bad that it is so difficult, as with its silvery foliage and attractive flowers, the prostrate forms would make excellent ground cover plants. The Oahu Island form is of easier cultivation and may well find its place in landscape work in the future.

Camphusia glabra: This beautiful plant only causes frustration; cuttings do not root and seeds do not grow.

This plant may be likened to a yellow-flowered Scaevola, and is often still known as such. If the secret of its cultivation can be found, it would be a most desirable plant for cultivation.

Only once did I get seeds to germinate; after several sowings involving over 100 seeds, over a period of three years, using all techniques that I could think of, two seedlings unexpectedly germinated. Ironically, no special treatment was used on the seeds. These seedlings refused to grow larger than two inches and after three months suddenly died. Cuttings have never shown any signs of rooting.

Currently seeds are being treated with Gibberelic acid, but

with no results to date. Our next attempt will be to graft it onto
stocks of the common "Naupaka", Scaevola taccada. As with all
Hawaiian plants which "cannot be grown", there has to be a way,
as occasional seedlings are still to be found in the wild.

Hibiscadelphus hualalaiensis and H. giffardianus: Generally
speaking, these members of the Malvaceae are of easy culture in
the propagation stages. Seeds usually grow well; though cuttings
have not rooted for me. It is important to use fresh seeds.

Being mid-elevation plants, perhaps the only way to grow them
successfully in the lower elevations is to grow them under about
25% shade.

Hibiscadelphus distans: This species, recently discovered in
Waimea Valley, Kauai, by Dr Herbst, is a dryland, "low" elevation
plant. To date it has proved to be of easy cultivation from seeds
and we have a plant flowering that is less than three years old
from seed. While not of spectacular beauty, it is of considerable
interest as the flowers are greenish and the young leaves are an
attractive silver colour.

Kokia drynarioides and K. kauaiensis: These plants are
similar to Hibiscadelphus hualalaiensis in that they are of easy
cultivation from seed and can probably be grown in the lower
elevations, as they grow fully exposed in the wild at
c. 1500-1900 ft. elevation.

K. cookei, reduced to a single plant in cultivation on
Molokai Island, it has been saved from extinction by the Cooke
family. Apparently short-lived, it is from quite a dry area of
Molokai, but the remaining plant is in a moister area and the
plant is declining. Fresh seeds germinate readily and at one time
I had eight plants growing on Kauai. On a recent visit I noticed
that they apparently had died. The problem is getting fresh
seeds and we are now working on getting the owner to send over
fruit as they mature. We also plan to air-layer the plant.
Cuttings under mist and with heating cables fail to do more than
make a little callus and in the past have gradually died out.

It will be sad to loose this species, which has barely
managed to stave off extinction; the flowers are an attractive
red and as a small tree it is considerably attractive.

Euphorbia degeneri: Of the several endemic species of
Euphorbia which are endangered, this one is perhaps the strangest
because of its ease of cultivation. In the wild it lives on
extremely poor soil or in pure sand, while seedlings brought into
cultivation grow with an ease which is amazing! Cuttings root
readily and seeds will germinate. The form of the plant is con-

siderably different in cultivation, due to the abundance of rapid
growth that it makes.

REFERENCES

BISHOP, L. (1973). The Honolulu Botanic Gardens Inventory 1972.
     Honolulu, Hawaii.

DEGENER, O. (1975). Notes from Waimea Arboretum 2(1).

FOSBERG, F. and HERBST, D. (1975). Rare and endangered species of
     Hawaiian vascular plants. Allertonia 1: 1-72.

HIRANO, R. and NAGATA, K. (1972). A Checklist of indigenous and
     endemic plants of Hawaii in cultivation at the Harold H. Lyon
     Arboretum. University of Hawaii.

OBATA, J. (1967-1973). Newsletter of the Hawaiian Botanical Society
     6(3); 10(5); 12 (1 & 2).

ST. JOHN, H. (1973). List of flowering plants in Hawaii. Pacific
     Tropical Botanical Garden (Memoir No. 1).

DISCUSSION SECTION III TECHNIQUES OF CULTIVATION

Mr Sipkes (Rockanje, Netherlands) asked about the importance of day length for arctic plants in cultivation.

Professor Böcher (Copenhagen) said that special equipment simulating the midnight sun was very important in maintaining the plants in cultivation and gave some details of the apparatus that he is using.

Dr Stearn (British Museum, Natural History) asked for details of the hot water treatment of seeds described by Mr Woolliams.

Mr Woolliams (Hawaii) replied that there are two types of treatment. Certain Malvaceae respond to warm water treatment. The seeds are placed in water at about 70°C, which is then allowed to cool naturally. With some legumes the seeds are put into boiling water and similarly cooled.

Dr Hall (Cape Town) asked for details of the number of staff at Waimea and the finances of the garden.

Mr Woolliams (Hawaii) replied that the project was funded by a private corporation who have a special interest in plant and bird protection in the 1800 acre valley, and that the project had been under way for three years. There are three administrative and twenty general staff.

Dr Popenhoe (Miami) asked how important it was to duplicate their rare species with particular respect to the Pacific Tropical Botanic Garden which is also in Hawaii.

Mr Woolliams (Hawaii) said that because the Hawaiian flora is fragile and difficult to cultivate duplication is essential and ideally seed and plant material should be shared with other gardens, but in practice this did not always happen. The ultimate aim of the recently formed Council of Botanic Gardens and Arboreta was to share and eliminate unnecessary duplication.

Sir Otto Frankel (Canberra) said that it is essential that in the conservation of domesticated plants all collections should be duplicated. He asked whether there was a plan to return conserved plants to the wild when suitable numbers had been propagated.

Mr Woolliams (Hawaii) replied that they would be returned to
the wild under scientific control, but that problems would result
if gene pools were mixed. He hoped that the Council of Botanic
Gardens and scientists would provide some guidelines but his
immediate concern is how to grow them as there is insufficient
knowledge of propagation and lack of cultivation records. Every
effort should therefore be made to distribute information to others
and at all cost to keep plants alive.

Mr Coode (Kew) asked if there was good liaison between foresters
and botanic gardens to induce advance flowering or fruiting of mature
trees.

Mr Woolliams (Hawaii) doubted whether such investigations had
been done and suggested that botanic gardens would be well advised
to undertake such research themselves.

Dr Budowski (IUCN Switzerland) asked what scientific groups
such as the IUCN and this conference could do to help solve the
problem of damage by goats.

Mr Woolliams (Hawaii) emphasised that in the light of the
Smithsonian Report the Conference should pass a strong resolution
to impress legislators into taking some action. Goat hunters,
responsible for goat conservation, comprise less than 1 per cent of
the population but are nevertheless a very influential minority
lobby. It is our duty to prove to the hunters that the goats are
responsible for the extinction of many rare Hawaiian plants.

Mr Lucas (Kew), referring to Mr Coode's question, told the
delegates of a recent conference of the International Union of
Forestry Research Organisation (IUFRO) which dealt with the production of fertile juvenile 'orchard trees' by grafting scions from old
matured trees onto new rooting stocks. He also suggested that it
should be possible to graft one sex of rare and scattered dioecious
trees onto the other in order to produce seed.

Documentation

THE RECORDS SYSTEM OF UPPSALA BOTANIC GARDEN

Örjan Nilsson

Uppsala Botanic Garden

Sweden

A new record system for the plant collections of Uppsala Botanic Garden was introduced in 1971 to replace the system built up by my predecessor, the late Professor Nils Hylander. Thus it is rather recent and our experience is limited and at present only a small part of the plant material in the garden is recorded according to the new system. This reorganization system is being introduced area by area, and new plants recorded in succession. We hope that our entire plant collections will be recorded within a period of ten to fifteen years. This long period of time has been calculated with regard to our present resources.

The record system is not new but has been developed from the older systems used in our garden. In its planning much attention was paid to systems used in other botanical gardens; mainly Scandinavian. We have also taken into consideration whether it is really necessary to have a record system at all, and to assess the drawbacks in relation to the advantages and costs. We soon realised that some type of recording was absolutely necessary, otherwise we would lose our control over the material within a few years. Then we had to choose whether we should retain the old system, which functioned but had some defects, or design a new one. We took a middle course and modified the old records, according to our needs and in this way we could use much of the existing material and did not need to change the format too much, which saved time and work. In planning this we consulted our colleagues, students, teachers, scientists and public visitors, and from them we secured much valuable advice, but it seems to me that our own practical experience was the most important factor in the reorganization.

The aim was to construct a record system which functioned with

the least possible friction and at the same time satisfied all our
needs particularly in reliability. Furthermore it had to be as
cheap as possible. After some adjustments we have now a system
which is close to these aims.

The system includes five components, each of which can function
and be managed separately, but which all interact. The co-ordination
of these parts is perhaps the novelty of the system. The different
components are as follows:

1. For many years separate planted areas of the garden have
   been named, eg. the Scandinavian Rockgarden, the System of
   Perennials, the New Sand Area, the Rosa- or Berberis-
   plains and the Northern Wall. These areas are further subdivided as required into smaller units which are separately
   numbered, and mapped in detail. The site of every species
   or collection is drawn and marked out by its name or code
   number and sometimes the number of planted specimens is
   also given. In long straight beds the plants are listed
   in a similar way. The maps and lists are numbered and
   kept in covers, one cover for each of the larger areas. If
   plants are moved, new ones planted, or the beds rearranged,
   the maps are modified, supplemented or redrawn. The maps
   give the opportunity to very easily correct errors caused
   by misplacement or destruction of labels. They also help
   us to guide visitors to particular areas in the garden to
   find specific plants. From the maps we can furthermore
   see where gaps exist when we need to plant new material.
   By referring to these maps it is possible to identify
   suitable sites where particular plants have thrived in the
   past.

2. Every collection which is planted out in the garden has an
   index card. The cards are kept in boxes in alphabetical
   order according to the name of the genera. Guide cards
   indicate the genera and are marked with their name and code
   number. The index cards are typed according to a model.
   If the plant dies or is rejected the card is excluded from
   the main index and held in archives. We think that in the
   future these archives will provide useful information on
   plants which we have tried but which are difficult to
   cultivate in Uppsala. The reason for, and year of,
   exclusion are noted on the card.

   Our new index has existed for about ten years. From the
   start different types of plants had cards in different
   colours, one colour for perennials, one for annuals, one
   for greenhouse plants, etc. The original cards which
   recorded almost all plant material in the garden were
   modified and incorporated into the new index, which saved

valuable time. The text on the front of the index card
agrees with the label. On the back the scientific name
with author is given. Sometimes important synonyms are
added, in particular those used in catalogues, etc. On
the back we also note the origin of the collection, its
locality and collector or the original donor garden or
firm, etc., and the year of acquisition. We also note the
name which the plant had when obtained. Phenological
observations as well as damage by frost during hard winters
is recorded for some species. If the original collection
is replaced by later generations raised in the garden,
this is noted, as well as which generation it is. As a
rule the existing collections are never mixed with newly
accessioned material which is grown separately and
recorded on its own individual index card. All name changes,
planting sites, etc., are recorded and the cards corrected
and if necessary refiled in which case a card is left in
the index to provide a cross reference. The reverse side
of the index card is thus an important source of information.

3. The text of the card shows the plant family, its scientific
   name (but not the author) and Swedish name, its distribution
   and a code number. The latter consists of three groups of
   figures, eg. A112:12:4. The first number group indicates
   the genus, with an 'A' for all generic names commencing
   with this letter, eg. Alchemilla. The figure is allocated
   according to the order of registration of the material.
   We started with Anemone, which thus has the genus number
   'A1', but before allocating new numbers the index is
   checked to avoid duplication. At the end of each letter
   group in the index there is a card recording all numbers
   used in succession. The second figure group indicates the
   species and is chosen in the same way and the third figure
   indicates the collection. By using the example above it
   can be seen that we have Alchemilla glomerulans from four
   different localities. It also indicates different intra-
   specific taxa, from cultivars to subspecies. The annual
   plants are the exception to this system, their labels are
   similar but the last figure in the code number only refers
   to intraspecific taxa. The reason is that their origin
   often changes from year to year. A small wooden label at
   the base of the main label indicates their current origin.
   On annual plant index cards information is collected each
   year regarding origin, purity of the cultivars, seed
   viability, gardens or firms from which the material was
   obtained, etc. The code number system functions but is
   rather time consuming and we are now considering dropping
   it and to arrange our index alphabetically according to
   the scientific names of the plants. Old labels are rather
   inconsistent and their text usually shows only the

scientific name, origin and sometimes year of acquisition.
The new labels for outside plants are of a uniform shape
and size and are made of engraved laminated plastic. In
the beginning we used dark grey-green labels with a white
text. The colour was chosen mainly to make the appearance
of the labels as discreet as possible. However, it was
soon evident that the labels became difficult to read as
the colour faded and the white text was obscured by dirt.
We are now using almost white labels with a black text,
and these seem to keep their colour better. After engrav-
ing, the plastic label is fixed in a frame and is attached
to a steel shaft of varying length, depending on the size
of the plant material. In the glasshouses we use other
types of labels but made of the same material. The label
remains with the plant throughout its life in the garden.
If a plant dies or is rejected the label is taken in. The
index card is removed from the index and site maps amended.

4. The nursery is one of its most important sections of the
   garden particularly as we have a special interest in annuals
   All new plant material enters the garden through the nursery
   including seeds, plants and cuttings. The nursery is
   responsible for propagating our own material and is also a
   research area. During their time in the nursery plants
   have wooden name labels with a number which is changed
   every year and this indicates where the plant is located,
   the origin of the plant is also shown. This date is
   recorded in a nursery index which is updated each year.
   The nursery tests perennials for hardiness, the name of
   the plant is determined, its place in the garden decided,
   and an index card prepared. After registration and
   labelling the plants are finally ready for the garden.
   The plants for the glasshouses are treated in a similar
   way. No plant is planted out in the garden before passing
   through the nursery.

5. Voucher specimens are obtained from plants growing in the
   collections and filed for reference. Each specimen is
   labelled with the text from the reverse of the index card.
   The dried specimens are held in a separate garden herbarium
   and if possible, a seed sample is also taken for the seed
   collection. This system has been in use in Uppsala for a
   long time and is a very valuable source of information.

   To date our experience of the system is good but to start with
it is rather time consuming and requires much care. In Uppsala we
often have a long winter which can be used for much of the record
work. It has been calculated that about 700 species may be recorded
per year by two persons in two months. However, because of limited
funds and shortage of staff we have only been able to pursue the

work intermittently. The cost of the system is therefore difficult
to calculate, but according to our present experience it is more
expensive than the previous methods. Because of lack of time and
personnel some parts of the system have been neglected. The system
is entirely manual but can, if we find it necessary, be partly
computerised. I am sure that the system can be adjusted and
rationalized at several points but at present we are confronted
with more serious problems. The system is adapted to meet the requirements of a medium sized botanic garden of a university. We
have at present about 10,000 species. Our small nursery has
become something of a stumbling block where at times plant material
tends to accumulate waiting for flowering to determine identity.
This has been balanced by a reduction of the accessions during
recent years. The advantage of the system is that it means
diversified security and reliability in the long run, good order
and control over the material which is for the benefit of all those
working with, using, or enjoying the plants of our garden.

THE USE OF RECORDS SYSTEMS IN THE PLANNING OF BOTANIC GARDEN COLLECTIONS

J. Cullen

Royal Botanic Garden, Edinburgh

Scotland

The idea that the collection maintained in a Botanic Garden should be planned is, in general, a relatively recent one. While it is true that certain gardens have, from the outset, had guide-lines as to what they should contain (such as plants from distinct politico-geographical regions, eg. Rancho Santa Ana and the South African botanic gardens) it is also true that the collections in the great majority of the European gardens have 'just grown', without any very precise policy for the acquisition and maintenance of stocks. The collections in such gardens have been highly selected; but the selection has been, by and large, outside of the control of those running the garden. Many factors have operated in this selection, some determined by the site of the garden, others more random; the following seem to be the most important:-

1. the range of climates available (ie the climate of the site and the provision of glasshouses, etc.);

2. the size and efficacy of the staff involved in maintenance;

3. the nature of the material available, whether from expeditions, the trade, or exchange between gardens;

4. the ease of propagation of the material available (selection by ease of propagation being a major influence on what material is widely exchanged between gardens); and

5. the purpose(s) served by the garden: amenity, research, education and conservation, or any combination from among these four.

Dr H R Fletcher, formerly Regius Keeper of the Royal Botanic Garden, Edinburgh has provided us with a vivid illustration of how this selection operated at its worst. Addressing a meeting celebrating the 150th Anniversary of the Botanic Garden at Geneva (Fletcher, 1969) he discussed the early introduction to European gardens of species from China:

"During the last twenty years or so of the last century, considerable amounts of seed were sent to some European gardens, from Western China, by the French missionaries Delavay, Soulie, David and Farges (selection by the collector). The seeds were sown in the usual fashion in the usual composts, and if and when they germinated the young plants were given hot-house treatment - and most of them promptly died ..... the gardeners simply treated the seeds and seedlings as they were accustomed to treating seeds and seedlings, and as they thought best (selection by staff expertise) .... During first half of this century pounds and pounds of seeds from the Himalaya were sent to botanic gardens in Britain - mostly to Kew and to Edinburgh - by E H Wilson, G Forrest, F Kingdon-Ward, R Farrer, J Rock, G Sheriff, F Ludlow, George Taylor, and others again, selection by collector - many of these collectors worked for horticultural syndicates who were interested in novelties and amenity plants- and by the purpose of the garden. Some of the seed was sown in botanic gardens and some of it distributed from botanic gardens to private gardeners and nurserymen (selection by exchange) - and sown by them . ....

Profiting from the mistakes of the early sendings of the French missionaries, if and when the seeds germinated, the seedlings were given cooler conditions, and thousands of young plants were raised .... . Horticulturists and gardeners did as they thought best; plants grown from seeds collected from above 11-12000 feet (3350-3650 m) were planted out of doors, sometimes in different parts of the garden. Sometimes the plants lived and sometimes they died (selection by climate and staff expertise)."

This is certainly a very bleak and perhaps extreme picture of the very unscientific (and unplanned) way in which collections were accumulated, but its general application to the whole range of collection development as it has been carried out heretofore is unquestionable. The net result has been the establishment of the notorious 'standard botanic garden collection' - easily propagated clones or lines of extremely robust, or extremely tolerant species - which forms the backbone, as it were, of the holdings of many gardens.

Such ad hoc, unscientific methods of acquisition and collection management are no longer tolerable; the input of expertise and energy (particularly for material under glass) per plant is too high for us to be able to continue with such a random methodology. In the future botanic garden collections must be planned both within and between gardens so that the best use can be made of scarce and diminishing resources. This planning depends entirely on the existence of adequate and accurate records.

The records system at Edinburgh has been comprehensively described elsewhere (Cullen, 1975) so I shall abjure any detailed account of it. It is sufficient to say that it provides listings of the gardens holdings arranged alphabetically by genera, by families, by collector (if known), by garden location, and by geographical origin. On the basis of these print-outs a catalogue was prepared, and distributed to most other botanic gardens in 1974. The preparation of such a catalogue is a step that I would like to recommend to other botanic gardens. Even in those cases where the level and standard of identification is low, a catalogue is of considerable help to all research workers. I know of about seventy such catalogues published since 1945, and a list of them is appended to this paper (pp.105-111). The consolidation of these lists, even though they vary widely in format would be a relatively simple clerical job, and would provide an extremely useful listing of what species are available in cultivation. Its value would be reduced, though not entirely vitiated, by the varying standards of accuracy of identification.

In order to plan a collection two factors have to be taken into consideration: (i) the environmental and staffing constraints; and (ii) the purpose or purposes which the garden has. This second factor will vary from organisation to organisation, but at Edinburgh, as at Kew and many other national or university gardens open to the public and allied to a research institution, it is very complex, and may be broken down into several different aspects.

1. amenity;

2. direct education (the provision of material for classes in botany and horticulture at various levels);

3. indirect education (material used in making the general public aware of the importance and fascination of plant life);

4. direct research (provision of material for research in progress in the related institution);

5. indirect research (the holding of a reservoir of species which can be made available to researchers in various disciplines);

6. conservation (a relatively recent development).

In any particular garden these different functions will be variously emphasised, and varying priorities will be given to them.

Planning must take account of, and indeed, try to harmonise these varying emphases.

At Edinburgh the policy of collection planning has now begun, using the facilities provided by the record system as the basic data. The overall plan is taxonomically based, using families (and to some extent genera) as the working units. These are assessed in terms of the environmental and staffing constraints and the purposes of the garden, as listed above, and each is given a priority level (one of five possible levels - see below).

The highest level of priority comprises those families and genera that are required for the purposes of direct research and direct teaching within the garden. Currently, research is being carried out on the Ericales (Ericaceae, Epacridaceae, Empetraceae and Diapensiaceae), Gesneriaceae (in particular the Old World representatives), Zingiberaceae; Orchidaceae (mainly from SE Asia), primitive woody Dicotyledones and Coniferae as well as on some individual genera from other families. In these cases we are trying to obtain as much material as possible, preferably of wild origin, and to replace stocks of unknown origin by fully recorded material. It may be said that in this category we are still operating the policy of growing everything we can get.

These collections are, of will be, one of the bases for published research, and so will have to be retained after the research itself is completed. It is important that such collections should not be broken up at the conclusion of active research; this is a very long-term commitment. Also included in this category are species that are used regularly in the teaching programme of the garden mainly classes for the Edinburgh Diploma in Horticulture, and in the exhibits available to the public in the Exhibition Hall. All these groups are clearly defined, and present no problems in terms of selection of material.

Problems arise when we consider the functions 3. and 5. - the indirect research and education functions. Both of these depend on the garden holding what has often been called ' a general representation of the world's flora' - a difficult concept which I have not seen analysed anywhere. There has been an unspoken tendency to assume that the mixture produced by the ad hoc methods described at the beginning of this paper was the best approximation possible: this, in fact, was often a cover for growing purely amenity material, and certainly cannot stand up to scientific scrutiny.

Under ideal conditions, a 'general representation of the world's flora' might well be a defined proportion (measured generically and specifically) of all taxa; but, of course, circumstances are not ideal, and such a simple measure is impossible. Some of the problems that produce this 'impossibility' are discussed below:

1. Many taxa are simply not obtainable. While it may be possible
to ask expeditions to look out for particular species, there is no
great likelihood that suitable propagules may be found. The expedition may well encounter the plant at an unsuitable time for the
collection of seeds or cuttings, or, if potentially suitable cutting
material is present, the area in which it is found may be so remote
that it is impossible to transmit the material in time for adequate
rooting. Even if suitable material is received at the garden, there
is no guarantee that mature plants will result.

2. Many species, though obtainable from time to time are impossible or extremely difficult to grow; many of the parasitic and sapro-
phytic species fall into this category, as well as species whose
growth requirements (climatic, edaphic or ecological) are difficult
to reproduce. In a large, general collection, such plants are not
worth the effort that must be put into their cultivation; but perhaps
in smaller collections, or where there is a research interest in
them, the extra effort can be applied, and successful cultivation
can be the result (Yeo, 1964).

3. A large number of species raise particular problems due to
their size and climatic requirements. The large rain forest trees
of the tropics fall into this category: no botanic garden in temperate regions can hold a comprehensive collection of,.for example,
Dipterocarpaceae. The same applies to a great deal of the purely
tropical flora; glasshouse space is usually restricted, and imposes
very severe limitations on the variety of species that can be grown.

4. A problem of a rather different nature is raised by the lack
of knowledge of the propagation and culture techniques required by
many species. The horticultural literature covers only a highly
selected and very limited range of plants, and the accuracy and
efficiency of many of the procedures recommended is, to say the
least, dubious. More research is an obvious answer to such problems;
but because of economic considerations, and the lack of suitably
trained staff to undertake it, such research seems unlikely to be
carried out. Over the years, however, garden staffs have developed
reasonably successful techniques of propagation and cultivation for
many species. These unfortunately are rarely published and are
passed on as a hidden oral tradition; many procedures die with their
inventor, and are lost. To prevent this happening, growers must be
encouraged to annotate their records with observations on cultural
techniques, even at the simplest level; this information can then be
stored, made available to others and considered critically, as and
when suitably trained staff are available.

The way we try to define 'a general representation of the
world's flora' has to take all these factors into account. Clearly
the level of representation regarded as adequate for a purely

tropical family will be lower than that for a temperate one; similar considerations apply to purely aquatic families, and others with a narrow range of ecological tolerance.

In trying to decide on levels of representation a number of particular problems present themselves:

1. What taxonomic rank should be the basic unit?

2. How can the level considered adequate be related to the size of the group in the wild?

At Edinburgh we have used the number of genera in the family as the working level; though not ideal, this is a convenient measure.   The representation of species within the genera is a matter for further discussion.   The relationship between the number of genera thought to give an adequate representation and the number of genera in the family is a complex one, not susceptible to precise definition.   In the larger families (ie those with more than twenty genera) we have considered ten per cent of the genera to be an adequate representation for temperate families, and four per cent for tropical or otherwise difficult families.   For smaller families the figure rises to a theoretical one hundred per cent for mono-generic families, whether tropical or temperate.

On this basis the families can be sorted into the priority levels mentioned previously:

1. The highest level, consisting of those families already discussed, of which we wish to achieve the highest level of representation possible.

2. The second level, consisting of families of which we require a better than adequate representation.   These comprise: (a) families related to those in the highest priority level, which may become important as research progresses; (b) families already well represented in the existing collection, for which expertise and suitable climatic conditions are already available; and (c) families which may well be used in direct research at Edinburgh in the future.

3. The third level, comprising those families which we regard as requiring an adequate level of representation - of no direct research interest but able to be grown in sufficient numbers.

4. The fourth level, comprising families of which we can hold only a token representation - again of no direct research interest, and unsuitable for large scale culture in

Edinburgh.   Many of the tropical woody families are included
here.

5. The lowest level, consisting of families that are impossible
   or extremely difficult to represent at all: parasites,
   saprophytes, and other difficult cases.

Some examples for levels 2-5 (level 1 has already been
discussed above) may serve to illustrate the use of these levels.

Level 2

Dilleniaceae - considered by some workers to be related to
the Ericales.
Betulaceae - although not directly related to the primitive
woody dicotyledons, this Amentiferous group (and others) may
have some bearing on the same problems.
Primulaceae - already well represented in the collection, and,
at least in part intensively studied at Edinburgh in the past.

Level 3

Caryophyllaceae, Leguminosae, Rubiaceae, Campanulaceae.

Level 4

Dipterocarpaceae, Meliaceae, Velloziaceae.

Level 5

Rafflesiaceae, Podostemaceae, Triuridaceae.

It must be emphasised that these levels are only guidelines to
be interpreted flexibly rather than rigidly.   Even so, they do
provide guidance as to what material is acceptable, and as to how
resources should be allocated.

There still remain problems of generic and specific representation.   Having decided, say, that we require a ten per cent level
of representation of the genera of the Rubiaceae, further guidance
is required as to which genera these should be, and which species
of these genera should be used.   There are many varying problems
like this, and we have decided to deal with them by means of small
committees of both scientific and horticultural staff looking at the
holding of particular groups.   A number of these groups are now at
work, on the conifers, palms, Orchidaceae, Betulaceae, and others
will be set up in the future.   These groups make decisions on what
genera should be included in the collection and what species should
represent them, bearing in mind the guidelines mentioned above, the
availability of material, and the morphological and geographical

range of the groups under consideration.

So far, I have not discussed the aspect of amenity, mentioned earlier as a function of a botanic garden open to the public. In general, it is possible to grow the plants which have been acquired under a planned policy in a way which is usually attractive, and displays the individual specimens to best effect. Some plantings, such as shelter belts, may have to be of species which are selected purely because of their ability to grow in the particular site where they are needed, rather than because of their specific importance in a planned collection; but these are necessary for the successful growing of other plants, and therefore cannot be said to be pure amenity plantings. Another possibility is to consider certain parts of the garden as purely amenity areas, such as bedding schemes around buildings, or heather gardens, etc. The total separation of amenity plantings from the broadly scientific collections is, however, not altogether a good thing (even though intensively worked collections may have to be segregated for some time), because, by adopting it, we lose the opportunity for public education by stealth. In the pleasant, well-laid-out surroundings of a botanic garden, the public is, on the whole in a receptive mood; by careful planting and discreet labelling, certain important ideas can be presented:

1. the importance of plant life to human societies;

2. the practical need for conservation;

3. the range of plant life suitable for small, domestic gardens;

4. the types of garden treatment suitable for difficult sites - around buildings, walls, etc.

This kind of planning will enable us to develop the collection in a rational way, and to use resources of all kinds to better effect. A similar kind of planning on a national or international scale is a further desirable development. As mentioned earlier, however, such planning is only possible for gardens which have an adequate record system: planning on an international scale requires some form of agreement on compatibility of at least the basics of record systems. This could perhaps come about as a result of this conference.

REFERENCES

CULLEN, J. (1975).   The living plant record system at the Royal
    Botanic Garden, Edinburgh.   In BRENAN, J.P.M., ROSS, R. and
    WILLIAMS, J.T. (eds.) Computers in botanical collections:
    167-176.   Plenum Press, London, New York.

FLETCHER, H.R. (1969).   The botanic garden as an experimental
    station; from the collector to the horticulturist.
    Boissiera 14: 57-64.

YEO, P.F. (1964).   The growth of Euphrasia in cultivation.
    Watsonia 6: 1-24.

PRELIMINARY LIST OF BOTANIC GARDEN CATALOGUES, INVENTORIES, ETC.*

ADELAIDE (1965). University of Adelaide, Waite Agricultural
   Research Institute, The Waite Arboretum. List alphabetical,
   latin names. Anonymous, duplicated.

ADELAIDE (1958). The Botanic Garden. Centenary volume, 1855-1955.
   History, Guides and Catalogue of Plants. List alphabetical,
   latin names, garden location, other information. Anonymous,
   printed.

AKUREYRI (1966). Listi Yfir Plontur Lysligards Akureyrar. Photos,
   map. 3 lists: herbaceous exotics, woody exotics, natives,
   systematically arranged, latin names. Anonymous, printed.

ALGIERS (1952). Le Jardin d'essai du Hamma. Photos, map. List
   of families alphabetically, latin names, notes. P. Carra
   and M. Gueit, printed.

ATAGAWA (1972). A List of Plants in Atagawa Tropical Garden and
   Alligator Farm. Photos. List systematic, latin names,
   authorities. Printed.

BERKELEY (1961). Trees of the Eddy Arboretum. Maps. Separate
   lists of Pines, Pine hybrids, Taxads and other conifers,
   alphabetical, latin names, authorities, notes.
   A.R. Liddicoet & F.I. Righter, USDA Forestry Service,
   Miscellaneous paper No. 43, ed. 2, duplicated.

BERLIN (not dated). Dendrologischer Führer durch das Arboretum
   des Museums für Naturkunde an der Humboldt Universität zu
   Berlin. Photos, maps. List alphabetical, latin names,
   notes. W. Vent & D. Benkert, printed.

BOGOR (1957). An alphabetical list of Plant Species cultivated
   in the Hortus Botanicus Bogoriensis. Photos, map. List
   alphabetical, latin names. P.M.V. Dokkus edited by W.M.D.
   van Leeuwen, printed.

BRADFORD (1969). The University Botanic Garden, List of Growing
   Plants. List systematic, latin names with authorities
   (includes Bryophytes). Anonymous, duplicated.

BREMEN (undated). Wegweiser durch den Rhododendron Park.
   Photos, map. Various lists: Rhododendron hybrids, Azaleas,
   Rhododendron species, evergreen shrubs, latin names.
   Anonymous, printed.

BRISBANE (1962). Catalogue of Plants. List alphabetical, latin
   names. Anonymous, printed.

BRNO (1971). Arboretum of the College of Forestry, Brno. Photos.
List (conifers and flowering plants separate) alphabetical.
Anonymous, printed.

CADENABBIA (undated). Vile Carlotta, <u>Elenco delle Piante</u>. Photos.
List alphabetical, latin names, notes. Anonymous, printed.

DAKAR (1953). <u>Le Parc Forestier et Zoologique de Hann</u>. Photos,
map. List alphabetical by families, latin names, notes.
J.G. Adam. printed, with supplement of additions for 1954
and 1955.

DARWIN (1974). Darwin Botanic Garden. List systematic, latin
names with authorities, notes. Anonymous, duplicated.

DAWYCK (1966). Dawyck Gardens, Stobo, Peeblesshire. <u>Trees at
Dawyck</u>, 1961 and 1966. List alphabetical, latin names.
Anonymous, duplicated.

DEHRA DUN (1954). <u>A list of Plants grown in the Arboretum and
Botanical Garden of the Forest Research Institute</u>. List
alphabetical, latin names with authorities, some notes.
M.B. Raizada and G.R. Hingorani, printed.

EDINBURGH (1974). <u>Catalogue of Plants in the Royal Botanic Garden</u>.
List systematic, latin names. Anonymous, printed.

EXETER (undated). <u>Catalogue of Coniferales on the University
Estate</u>. Map. List systematic, latin names with authorities.
Anonymous, printed.

FUNCHAL (1966). <u>Some Trees, Shrub and Plants of Interest grown at
"Quinta do Palheira" Madeira at 1700 ft. and also in Town
Garden of "Quinta Santa Luzia" in Funchal at 450 ft</u>. List
alphabetical, latin names. M.A. Blandy, duplicated.

GLASNEVIN (1960). <u>Hand List of Orchids cultivated in the National
Botanic Garden</u>. List alphabetical, latin names with authorities. Anonymous, duplicated.

GOTEBORG (1963). <u>Katalog over Vaxterna i Goteborgs Tradgard, 1962</u>.
Photos, map. List alphabetical latin names with authorities;
also separate lists of wild plants. B. Lindquist, printed.

GROENENDAL-HOEILAART (1955). <u>Catalogue de l'arboretum de Groenendal</u>
Photos, maps, tables. List alphabetical, latin names, many
notes. Anonymous, printed. (Station de Recherche de Groenendal
Trav., Ser. B, No. 19).

HAMILTON (1969). Checklist of Spontaneous Vascular Plants of the
    Royal Botanic Garden. Map, photos. List systematic, latin
    names. J.S. Pringle, printed. (Royal Botanic Garden Technical
    Bulletin No. 4).

HAWAII (1970 and 1972). A Checklist of Palms in the Harold L. Lyon
    Arboretum, University of Hawaii; A checklist of Ficus ...
    Lists alphabetical, latin names. R.T. Hirano and K.M. Nagata,
    printed.

HELSINGBORG (1964). Botaniska Tradgarden inom Frederikodalo
    Friluftsmuseum. Photos, map. List alphabetical, latin names.
    H. Vahlin, printed.

HIDCOTE (1963). Guide to Hidcote Manor Gardens, including list of
    principal trees and shrubs. List arranged by garden location,
    latin names. Anonymous (National Trust), printed.

HONOLULU (1972). Honolulu Botanic Garden - Inventory. Photos.
    List systematic, latin names with authorities. Anonymous,
    printed.

HOYT ARBORETUM (1974). Hoyt Arboretum, Portland, Oregon. List of
    conifers, alphabetical, latin names, notes. Anonymous,
    duplicated.

KAMON (Szombathely) (1956). A Kamoni Arboretum. List alphabetical,
    latin names, notes. J. Bano & J. Retkes, printed.

KANAGAWA (1972). Kanagawa Prefectural Ofuna Botanical Garden,
    List of Plants Cultivated. List systematic, latin names
    with authorities, notes. Anonymous, printed.

KEW (1962). Handlist of Orchids in the Royal Botanic Gardens, ed. 3.
    List alphabetical, latin names, with authorities. Anonymous,
    printed.

KIEV (1970). Botanichnii Sad im Akad. O.V. Fomina. Photos, map.
    List systematic, latin names, notes. Anonymous, printed.

KISHINEV (1967). Parki Moldavii. Photos, map. List alphabetical,
    latin names in tabular arrangement, notes. P.V. Leont'ev,
    printed.

KOBE (1968). Kobe Municipal Arboretum. List of Trees and Shrubs.
    Map, photos. List systematic, latin names with authorities.
    Anonymous, duplicated.

KYOTO (1968). Kyoto Herbal Garden, Research and Development
     Division, Takeda Chemical Industries. List of Plants. Map,
     photos. List alphabetical, latin names with authorities.
     Anonymous, printed, (3rd edition).

KYOTO (1963). Nippon Shinyaker Institute for Botanical Research.
     List systematic, latin names with authorities. Anonymous,
     duplicated.

KYSIHYBEL (1960). Lesnicke Arboretum v Kysihybli. Photos. List
     alphabetical, latin names, extensive notes. M. Holubcik,
     printed.

LAL-BAGH (1969). Plant Wealth of Lal-bagh. List alphabetical,
     latin names with authorities, notes. M.H. Mari Gowda and
     M. Krishnaswamy, printed.

LENINGRAD (1961). Akad. Nauk. SSSR, Botanicheskii Institut inom.
     V.L. Komarova, Putevoditel's po Parku Botanicheskogs Instituta.
     Maps, photos. List alphabetical, latin names, notes.
     B.N. Zamyatnin, printed.

LIVERPOOL (various dates). Botanic Gardens, Harthill. Several
     lists, alphabetical, latin names. Anonymous, printed.

LJUBLJANA (1956). Arboretum Volcji Potok. Photos, map. 3 lists.
     Gymnosperms, Angiosperms (woody) and Angiosperms (herbaceous),
     alphabetical, latin names, notes. C. Jeglic, printed.

LONGWOOD (1970). Plants growing in Conservatories and Gardens,
     October 1970. List systematic, latin names. Anonymous,
     printed.

MAYAGUEZ (1967). Federal Experiment Station, U.S.D.A., Mayaguez,
     Puerto Rico, Plant Inventory 1967. List alphabetical, latin
     names, notes. Anonymous, duplicated.

MERION STATION (1970). Checklist of the woody plants at the
     Arboretum of the Barnes Foundation. List systematic, latin
     names. H.B. and J.M. Fogg, printed.

MIAMI (1949). Fairchild Tropical Garden, Catalog of the Plants
     in the Garden. List alphabetical, latin names, notes.
     W.L. Philips, printed, revised 1950.

NESS (1969). List of Plants cultivated in Liverpool University
     Botanic Gardens. List alphabetical, latin names with
     authorities. Anonymous, duplicated.

NEWBY HALL (UK) (1968). <u>Catalogue of some of the interesting Plants</u>
<u>grown in the Gardens</u>. List alphabetical, latin names.
Anonymous, printed.

OAKLAND (1967). <u>Trees and shrub of Miles College</u>. List alphabetical, latin names, notes. B. Kasapligil, printed.

OITA-KEN (1963). <u>List of Plants cultivated in the Oita Research</u>
<u>Station for Agricultural Utilization of Hotspring</u>. List
systematic, latin names with authorities. Anonymous, printed.

OLDENZAAL-DE-LUTTE (undated). Arboretum Poort-Bulten, Oldenzaal,
Holland. List of Plants. Map. List systematic, latin names.
Anonymous, printed.

OSAKA (1960). <u>List of Plants cultivated or wild in the Botanical</u>
<u>Garden of Osaka City University</u>. Map. List systematic,
latin names with authorities. Anonymous, printed.

OSAKA (1964). <u>Tennoji Park, List of Plants in the Greenhouse</u>.
List systematic, latin names. Anonymous, printed.

OSLO (1966). <u>Katalog over Plantene i Botanisk Have, Universitet</u>
<u>Oslo i aret 1964</u>. Map, photos. List systematic, latin
names with authorities. Anonymous, printed.

PALLANZA (1964). <u>Catalogue of the Plants, Villa Taranto Botanic</u>
<u>Gardens</u>. Map, photos. List alphabetical, latin names with
authorities. N. McEachern, printed.

REA (nr Turin) (1974). <u>Index Plantarum</u>. List alphabetical, latin
names with authorities. Anonymous, duplicated.

RIGA (1972). Rayon Stucka, <u>Skriverskii Dendrarii</u>. Map, photos.
Two lists, one tabulated, alphabetical, latin names.
A.V. Zvirgzd, A.M. Maurin <u>and</u> P.E. Tsinovskis, printed.

RIO DE JANEIRO (1972). <u>Indice das principais especies e generos</u>.
Map, photos. List systematic (others in temporal order),
latin names. Anonymous, printed.

ROGOW (1966). <u>Arboretum w Rogowie</u>. Map, photos. List alphabetical, latin names with authorities. H. Kryczynska, printed.

SAN FRANCISCO (1966). <u>Guide List to Plants, Strybing Arboretum</u>.
Map. List alphabetical, latin names, notes. E. McClintock,
printed.

SAN FRANCISCO (1966). <u>Trees of the Panhandle, Golden Gate Park</u>.
     Photos. List arranged according to Garden locations, latin
     names, notes. <u>E. McLintock and V. Moore</u>, printed.

SEATTLE (1969). <u>University of Washington Arboretum: a show case
     for all seasons</u>. Map, photos. Lists (2) alphabetical latin
     names and alphabetical vernacular names. Anonymous, printed.

SEATTLE (1969). Garden of the Hiram P. Chittenden Locks, Lake
     Washington Ship Canal. Map. List arranged according to
     location, latin and vernacular names. Anonymous, printed.

STANFORD (undated). <u>The Gymnosperms growing on the Grounds of
     Stanford University</u>. List systematic, with short descriptions
     and keys. Anonymous, printed.

WARSAW (1968). <u>Wykaz grintowich roslin zielnych i roslin
     szklarnowych</u>. Photos. List systematic, latin names with
     authorities. L. Karpowiczowa, printed.

WEIHENSTEPHAN (1971). Sichtungsgarten Weihenstephan. Map, photos.
     List alphabetical, latin names with authorities. R. Hansen,
     printed.

WELLINGTON (N.Z.) (1967). <u>Guide list to Plants in the Otavi
     open-air Native Plant Museum</u>. Photos. List alphabetical,
     latin names. Anonymous, printed.

WORLITZ (undated). <u>Bäume und Sträucher in Worlitzer Park</u>.
     Maps, photos. Various alphabetical lists, latin names, notes.
     K. Levins, printed, 2nd edition.

ZAGREB (1963). <u>Botanicki Vrt u Zagrebu Vodic</u>. Maps, photos.
     List arranged according to garden locations, latin names.
     S. Ungar, printed.

*The form and style of these publications is variable; many of
them seem not to have a recognisable title. Titles, when known,
are underlined. The lists are arranged under the name of the town
in which the garden occurs.

The following additional gardens list a catalogue among their
publications in the <u>International Directory of Botanic Gardens II</u>.

BRUSSELS, Arboretum de Tervuren.
CINCINNATI, Eden Park Conservatory.
CLERMONT, Bernheim Forest Arboretum.
EALA (Congo)
ENGLEFIELD GREEN, University of London Botanical Supply Unit.
HILLSDALE, Slayton Arboretum.
HYDERABAD, Botanic Garden.
KINGSTON, Royal Botanic Garden.
KOBE, Rokko Alpine Botanical Garden.
LAE, Botanic Garden.
NAPLES (Florida), Caribbean Garden (Bromeliads only).
NOGENT-SUR-VERNISSON, Arboretum Vilmorinianum.
SANGERHAUSEN, Rosarium (Roses only).
SAPPORO, Botanic Garden.
SENDAI, Botanic Garden.
WURM, Pforzheimer Alpengarten.
ZURICH, Staedtische Succulentensammlung.

THE FUTURE AND ELECTRONIC DATA PROCESSING

F. H. Perring

Biological Records Centre

Institute of Terrestrial Ecology, Huntingdon, England

The choice of an Electronic Data Processing system, or the decision whether to use one or not, should not be made until we know what demands will be made on the data. I am constantly reminded that the price of computing is rising; I never need reminding that staff salaries are also rising so before we embark it is more than ever essential to know we can afford what we shall use.

At BRC we have set up a modest information retrieval system for botanic gardens. As part of a plan for the conservation of the British Flora we thought we should find out which of our rare plants of known origin were already in cultivation in botanic gardens in the British Isles for two reasons.

1. As a stimulus to gardens to contribute to conservation by acquiring plants not already in cultivation in other gardens.

2. As a guide to experimental taxonomists of where material is available, and as a hint as it were, not to collect in the field when this is unnecessary. Further stimulus to seek material from gardens rather than from the field should come from the Conservation of Wild Creatures and Wild Plants Act which became law at the beginning of August 1975, under which it is now an offence to collect any material of twenty-one endangered species. Though a licence may be obtained by individuals for scientific purposes it will surely be easier, quicker and safer if gardens receive licences to collect and individuals acquire without licence their material from gardens.

113

To start this system off we have used a punched card design and from a simple form or from letters prepared a card for each plant reported to us as growing in the gardens which have corresponded. This enables us to print out a document for circulation which is an alphabetical list of species, their geographic origin, the garden in which they grow - and the availability of material in that garden.

Circulation of the first, very incomplete, list in 1972 based on returns from about twenty gardens, was a tremendous stimulus and the second list sent out in August 1974, was twice as long and had contributions from twice as many gardens. A third list is now in preparation and this looks like being three times the length of the first.

When we send the list out it has an appendix of desiderata attached - so far it has included only a general request to acquire material - though suggesting this mainly be confined to taxa in the "area" of the garden. Now that the number of taxa still to be collected is much smaller it may be wiser to ask particular gardens to be responsible for much more restricted areas.

As the system has now proved its usefulness and gardens are clearly willing to co-operate in this restricted exercise, we can perhaps justify the extra input of men and machinery which will be required to computerise the system to handle the increased number of records involved. So far we have only been considering about 300 species, but if we were to expand to cover critical genera and taxa below the species level several thousand would be involved, a card system would be clumsy and inappropriate and we would need to turn to EDP.

In thinking about the most suitable system to adopt I have leaned heavily on the report made here two years ago by R.P. Brown on the American Horticultural Society Plant Records Centre and on recent investigations made here at Kew - as well as on our own experience at BRC.

If the simple output I have described for our rare species survey is adequate then a system can be devised which involves a number of centres in data processing - similar to the Central Data Bank set up in the United States - which would act for all the other gardens in a country or other suitable regions (perhaps related to a limited number of phytogeographical regions). Each garden in the region which contributes would keep its own documentation and agree to send a copy of the documents to the regional centre. There is no reason why some large comprehensive gardens should not contribute to several regional data banks.

The basic information required for each record would be the

garden code, the name of the taxon, the authority, source of material, its geographical origin, date of receipt and its location in the garden. A standard data sheet used by all gardens would be prepared which contained space for all this information about each specimen needed by the garden and for all the cultural information, the importance of which has been referred to by many speakers. The sheets would be completed in triplicate.

1. Sorted to provide a species index for the garden

2. Sorted to provide a location index for the garden

3. Sent to the regional data bank when the plant is established in the garden.

The regional data bank would run an information retrieval system based on the use of computers and microfiche. Upon receipt each original document would have a unique number boldly and clearly added to it. The document would then be microfilmed and this number would become the identifier of the record in the system.

The original document would then be processed for computer store into which only the taxonomic name, authority, country of origin, garden in which growing, availability of material and the microfilm (document) number are entered.

This would enable the regional data bank to prepare and send to each garden:

a. A master list taxonomically arranged of species in cultivation in the gardens in the region

b. A master list similarly arranged for that garden

c. A library of information on microfiche about each specimen which can be consulted by referring to the unique number on the master list - tagged to show which have cultural data for example.

On request the regional data bank could also prepare a list of all taxa available from particular countries either for one garden or for any group of gardens which would be invaluable for briefing expeditions. If for conservation purposes a group of gardens within a phytogeographical region has as an objective the cultivation of a specified list of taxa then the regional data bank can produce updated lists of desiderata from the group of gardens or for individual gardens.

The advantages of this system over a fully computerised one are:

1. Safety - all data are immediately filmed

2. Economy - the Plant Records Centre in the United States was charging its contributing gardens $80-$100 for each hundred records two years ago, a price well beyond the reach of most gardens

3. Accessibility - each garden, via a microfiche reader, has access to the complete data in the system - not just a computer print out digest - so if there is space on the original document for cultivation notes for example, these would be available to the microfiche reader.

Updating would be done by rewriting the record on a new record sheet giving it another unique number, microfilming it and updating the computer file for the appropriate changes (eg. taxonomic name, availability of material) and the change in unique number. Old information would remain in the microfiche library but, because its reference number would no longer be operative in the computer it would no longer appear in the computer print out and would never be referred to. Because of the minimal storage space required for microfiche this accession of "dead" material would not embarrass the user - there would be no need to update each garden's collection (though there would be dangers in browsing).

The regional data bank would also have to be informed of deaths; this could be most simply done by each garden periodically sending in a list of unique numbers based on a review of the garden's master list.

Another advantage of this combined microfiche and computer system is that it could be used even with gardens which already have their own extensive card index system. It does not matter whether these conform to any standardised format - as long as the basic information is available this can be picked up direct by operators if this is reasonably clear or via a data sheet if it is not.

If the card index of a particular garden were only organised by species and not by location in the garden then the regional data bank could provide a service by preparing one set of microfiche for that garden alone, organised to show what was growing in particular sections of their garden. Once this had been done though the garden could begin to operate the '3-copy' system. However, the garden would have to operate its own deaths system on these records.

CONCLUSION

From our own limited experience with the rare species list it may be best to start the sort of system I have been describing by co-ordinating the methods of a few large gardens - nevertheless consulting representatives of smaller gardens - to produce a master list based on the complete collections of each (or of course for some pre-defined segment of the collections).

These large gardens are likely to have the staff necessary to contribute to the system and are the most likely to benefit quickly from the facilities offered by the system - the result should be the largest list of taxa for the least effort.

Other smaller gardens which do not wish to contribute fully and neither feel the need nor have the finance for their own detailed internal documentation could be asked to send to the regional data bank only records additional to those included in the first list from the large gardens. This would also mean that it would mainly be the large gardens able to cope which would receive requests for material. Smaller gardens would only receive requests for what they held uniquely. And if and when these plants were established in one of the large gardens they could withdraw them from the list by declaring the species "dead" or "not available".

I must apologise if this paper leaves many ends untied; other pressures have prevented me from giving as much thought as I would have liked to the problem. However, I hope some parts at least may commend themselves; I think there may be advantages in a simple system which is practical to which all may contribute at the level of their financial ability as against a fully comprehensive on-line system which would probably only be available to a few larger, wealthier institutions.

INDEX SYSTEMS FOR SPECIAL COLLECTIONS AS ILLUSTRATED BY THE

ORCHIDACEAE

Matthias Hofmann

Konstanz

Federal German Republic

In his paper on the "Threatened Plants Committee" Professor
Heslop-Harrison (1974) emphasized that "for vegetation in general,
much the best prescription for survival is conservation in the
field .... But we know that the rate of diminution of some habitats
is now such that field-conservation of some taxa may soon become
impractical". Therefore one should use the potentialities of
botanic gardens as well as private and commercial collections for
the conservation of threatened plants through cultivation.

If we wish to do so, it is necessary to know which species are
already in cultivation and how frequently a particular species is
cultivated in order to assess rationally which species should be
provided with costly culture-places. With this information one can
prepare an index of cultivated plants that can be compared with a
list of threatened plants in order to determine which species ought
to be brought into living plant collections. With this knowledge
it will be possible to economically plan conservation-orientated
collections. Last but not least, collections with such rare
exquisite plants will arouse public interest.

With an index system for living plant collections a useful aid
will be had in solving the many nomenclatural problems; and one can
produce standardized index cards for each collection as a basis for
garden plant labels. Further, it will be possible to answer inquiries from collectors and other interested persons. The final
step would be to produce an "International Botanic Gardens Yearbook"
containing the names and synonyms of all species actually cultivated.

The effort needed to achieve these aims is not very large if
one utilizes the capabilities of electronic data processing. In

order to establish this fact, a program for the TR440-Computer-System (1973) has been written with some data obtained from botanic gardens, as well as private and commercial collections. The program can be applied to all plant-species, but only orchids have been considered in this case due to the effort in collecting and recording the data.

The program consists of two parts. The first part applies features of the TR440-software and can use such conversational time sharing terminals as teletype writers or alphanumeric display terminals. It allows for memory storage, erasing and updating of names and synonyms of species cultivated in a collection; and it can store, erase and update names and addresses of collectors. It would be useful if the program also handled information on a particular species, such as important properties of habitats, references to literature and pictures of the plant, also if seeds or plants are available from a collection.

The program allows the computer to answer the following inquiries:

- in which collections is a particular species to be found?

- what are the names of species cultivated in a particular collection?

- which collections are involved?

- what are the various names associated with a particular species?

It is possible to expand the storing and inquiry-answering part of the program so that questions can be answered such as "which warm-house orchid from Costa Rica is most frequently cultivated in British botanic gardens?"

The second part of the program is less convenient, but could easily run on other computers that have batch processing facilities only. The program is non-conversational and serves to store data, and in particular to produce statistical lists. As the programs are presently used for Orchidaceae only the results are entitled "Statistics on Cultivated Orchids" Hofmann (1975). One of the currently produced statistics is on frequently cultivated species. This statistic is of particular value to those who are interested in cultivating threatened or rare species in their own collections. A second and larger statistic, lists approximately 5,000 species, subspecies and varieties as well as some natural hybrids contained in the nearly 17,000 orchideous specimens included from the fifty-five collections recorded. In addition, 2,500 synonyms have been included. This booklet might be a first step on the way to an "International Botanic Gardens Yearbook".

It is not difficult to expand the computer program to handle further data so that a particular collecting - having submitted the names of its plants - can obtain the filing card information on each species containing the basic data required to write garden plant labels. Naturally this can be done only if a standardized filing card exists.

Programs used to solve such problems are typically called IS&R, ie. "Information Storage and Retrieval", and a detailed description of the presented program is available, as well as an issue of the "Statistics on Cultivated Orchids".

The opportunity to utilize the computer equipment necessary to produce the "Statistics on Cultivated Orchids" for a presentation was on a one time only basis, and the work itself carried out only to establish the practicality of the ideas described above.

It is desirable that this work be continued; for example, expanding the number of considered collections in order to make the statistics more representative. Or one could begin to collect data on succulents. To expand the program in order to store further data on a particular species would also be important; you would then be able to produce standardized filing cards and answer more detailed inquiries. The linkage to already existing small local data bases should be discussed. What is missing at the moment, however, is the financial backing to pursue this project from this rather simple presentation to worldwide usage, providing a continuing Botanic Data Base.

Finally, two proposals to resolve this problem:

1. To locate an already existing fund, or to obtain a grant, that would provide a sum to an institute that possesses a large-scale computer. This institute would then do the data collecting and programming for a Botanic Data Base as previously described. In particular it would produce and sell an "International Botanic Gardens Yearbook" and answer collectors' inquiries for a small fee. After a certain time - perhaps one year - the sum could be repaid to the fund.

2. All conservation-orientated and interested botanic gardens establish a fund from which the costs for programming, data collection, and delivery of information to members is paid to an institute that possesses a large-scale computer and is prepared to do this work. Non-members of the fund may use the Botanic Data Base for a nominal consideration.

REFERENCES

AUERBACH PUBLISHERS INC. (1973).   Auerbach EDP notebook/Europe,
   Segment A: Telefunken TR440.   Philadelphia, Pa.

HESLOP-HARRISON, J. (1974).   Postscript: the Threatened Plants
   Committee.   In HUNT, D.R. (ed.).   Succulents in peril.
   Supplement to IOS Bulletin 3(3): 30-32.

HOFMANN, M. (1975).   Eine Statistik kultivierter Orchideen.
   Die Orchidee, Hamburg 26(1): 30-33.

HOFMANN, M. (1975).   Einige häufig kultivierte Orchideenarten.
   Die Orchidee, Hamburg 26(3): 140-141.

DISCUSSION SECTION IV DOCUMENTATION

Dr Stearn (British Museum, Natural History) raised the question of loss, theft or misplacement of labels on plants and suggested that small tubes containing essential records could be buried in the soil close to the plant concerned.

Dr Williams (Birmingham) stated that the documentation system must be adequate for the institution, and that there must be a ready exchange of information. There should also be some onus on the institution to maintain whatever stocks had been recorded in central data banks. He mentioned some recent collaboration in the British Isles but stated also that collections of taxonomic importance had been lost following changes of staff.

Dr Y Heslop-Harrison (Kew) commented on Dr Cullen's mention of reduction of stocks to simplify collection and asked whether a policy of distribution of surplus material had been established.

Dr Cullen (Edinburgh) replied that such a policy was in operation at Edinburgh for orchids but not for other groups.

SECTION IV SUPPLEMENTARY COLLOQUY ON RECORDS SYSTEMS

Dr Hall (Cape Town) summarised the many comments that had been made by delegates on the importance of exchange of information and the need for some form of centralised pool for transfer of such information. He said that an information system (or systems) was needed for three kinds of information:

    (i) Holdings of living collections in botanic gardens (including seed lists).

    (ii) Holdings of rare and threatened or endangered species.

    (iii) Performance information and cultivation information.

Professor Heywood (Reading) proposed adding to seed lists notes on the availability of other propagules.

Dr Cullen (Edinburgh) suggested that as far as system (i) was concerned the system of the American Horticultural Society was not detailed enough. The information needed in a collaborative system is:

    (a) What material do we possess?

    (b) Where is it located?

    (c) Why do we have it?

However, it is the responsibility of individual gardens to produce their own lists.

Dr Popenoe (Miami) stated that systems (i) and (ii) could be handled by the American Horticultural Society's system but the information needed for item (iii) could be handled by horticultural or scientific journals where performance and cultivation information on propagules and seed availability could be stated clearly.

Dr Hofmann (Konstanz) proposed that when considering item (i) some ideas were needed for the smaller gardens. Records systems for these gardens should be cheap, but if possible computerisable. When considering items (ii) and (iii), general information on threatened species and performance in cultivation were necessary.

Mr Paterson (Chelsea) stated that if possible more separate sheets of paper should be avoided. The present seed list could, however, be presented in a much more productive fashion to include such items as propagation and cultivation methods.

Mr Evans (Edinburgh) suggested that seed lists will be necessary as it would be known from such a record system where the material can be obtained. A propagule would give rise to a plant more true to type than one raised from seed where cross-pollination may have occurred. In the computer system at Edinburgh garden holdings are listed alphabetically, plants are listed according to their families and according to their origin and collector. The exact location of any plant in the garden is also recorded.

Dr Popenoe (Miami) agreed that information on computerised records could also include information on whether the plant was threatened or endangered.

Professor Heywood (Reading) asked whether speakers were referring to one data centre or to numerous centres. He also referred to the large size and high cost of the printouts involved.

Dr Popenoe (Miami) replied that the printouts of the American Horticultural Society contained 6,000 records and cost $60 for the first copy but much less for duplicates. He considered it unlikely that a printout of the total holdings was needed, only one covering endangered and threatened species. He suggested that perhaps the IUCN could provide the funds for this programme.

Mr Evans (Edinburgh) stated that such a printout would be a help in researching on certain projects so that the location of any particular species could quickly be determined.

Mr Keesing (Kew) suggested that it is the service of botanic gardens to provide scientific research material. Therefore there is a need for information on all material held in botanic gardens to reduce the strain on wild material due to unnecessary collecting.

Mr Simmons (Kew) summarised by saying that all three points were included in the records system at Kew. Kew is fortunate in being plugged into a Ministry computer because cost is a major item in establishing a computerised system. Specific lists of taxa must be built up. The seed list is now outdated and computer records will in future be the basis for exchange activity. Kew is now at the stage of building up information and has started negotiating with the Forestry Commission and the National Trust for places such as Bedgebury Pinetum and Westonbirt to adopt similar systems. He stated that one basic simple system is needed with a key provided to give access to background information.

Mr Crosby (Leeds) was concerned about the idea of abandoning seed lists. He pointed out that a great deal of work is done by research students who need material in a hurry. With only print-outs being available as sources of material, some material required urgently for research may be only vegetative and might not be ready until too late.

Professor Heywood (Reading) seconded by Dr Popenoe, proposed the setting-up of a working party which should produce a report in a year on the feasibility of setting up an international reference system for botanic gardens. The working party to inspect any existing data systems and make recommendations.

Dr Cullen (Edinburgh) suggested that the TPC should be involved in the working party.

Dr Harris (Zaria, Nigeria) added that the possibility to expand to include all holdings should also be considered.

Dr Hall (Cape Town) said that the working group would have to report back to an executive representing the Conference, and that the report would need to be circulated to every member of the Conference.

Dr Harris (Zaria, Nigeria) suggested that the action should be based on Edinburgh, Kew and the American Horticultural Society, and this was agreed by Dr Popenoe.

Dr Hall (Cape Town) added that those three institutions would have to find representatives to act on the working group, and it was agreed that Dr Cullen should co-ordinate matters.

Dr Y Heslop-Harrison (Kew) spoke of the difficulty of deciding what was rare and asked if for example the Red Data Book should be used as background data.

Dr Harris (Zaria, Nigeria) mentioned the definition of terms previously given by Mr Lucas.

Conservation of
Horticultural Plants

C. D. Brickell

Royal Horticultural Society

Wisley, England

The need for the conservation of wild plant species is rightly being emphasized throughout the botanical world. Campaigns have been mounted, papers written, conferences and seminars organised and governments harangued in order to highlight the intense need for co-ordinated efforts to be made to achieve this aim. At the present moment the majority of effort in plant conservation is understandably concentrated on natural plant populations, but there is, in my view, an urgent need for similar conservation policies to be adopted for cultivated plants in the widest sense. There are enormous problems to overcome and there are in Biblical analogy the "doubting Thomases" who will consider any such schemes to be impractical.

One has only to consider the very similar difficulties which beset the conservation of wild plants and the strides made towards an integrated policy in a very few years to know that if the aims are worthwhile and the determination to succeed sufficient then good results can be achieved.

It is easy to state that Conservation with a capital 'C' is required in this, that or the other field, but justifying the need is as essential in this as in all other cases.

I would like to put forward for consideration the following points which are basically those which also apply to wild plants.

    1. Economically it is important to retain certain hybrids and cultivars which are of genetical value in breeding plants used in a variety of ways by human beings.

2. The loss of the genetic potential of plants cultivated and
   used by man, whether in farming, forestry, horticulture or
   medicine, increases as old cultivars are replaced and
   gradually die out through neglect.

   Economically this may prove disastrous if, for example, the
   potential for disease or pest resistance or the ability to
   withstand adverse conditions is not passed on to the offspring.

   The new cultivars may appear economically attractive
   initially, perhaps because of higher yield, or some other
   character, but the cultivar concerned may succumb to disease or some other trouble later. Breeders know only too
   well how this happens. Unfortunately, an industry may
   within a very short time turn over to the "new improved"
   cultivar which appears commercially the better plant. If
   the worst occurs, which may be through no fault of the
   breeder, and the "new improved" plant succumbs, then the
   problems in establishing breeding programmes to replace it
   are magnified considerably if no-one has bothered to retain
   the old cultivars with resistance potential which may have
   occurred far back in the ancestry of the "new improved"
   plant. Some old cultivars are irreplaceable if lost, as
   their parentage is unknown and it is unlikely that they can
   be easily recreated.

   This has been realised for a considerable time in agriculture and forestry, where for certain crops the conservation of important old cultivars is reasonably established.
   In horticulture this aspect is scarcely considered at present and there is no integrated conservation policy for
   cultivated plants. The need is there whether the plant
   concerned is used for food, medicine, clothing, ornamentally
   or in any of the numerous ways that Man exploits the plant
   world.

3. Historically it is important to maintain a record of Man's
   achievements in plant breeding whether accidental or inten-
   tional. We proudly maintain collections of an extraordinary range of objects associated with Man's history
   and yet give the minimum of attention to cultivated plants
   upon which the human race depends for survival.

   It might well be argued that the human race has got along
   fairly well without any conservation policy for cultivated
   plants with negligible ill-effects so far - so why bother
   now? The answer to this not unreasonable question is much
   the same as those which apply to our wild plants.

Population pressures and economic pressures are basic to the
problem. As the world population grows the demand for
plants for a multitude of purposes increases. New cultivars
are bred with increased yield, quality or other "improved"
characters and the old cultivars of no further apparent use
are neglected or destroyed. Economics dictates that most
nurserymen's lists decrease and gradually cultivars which
may be less prolific of increase or yield are disappearing.
Sadly even the EEC by producing lists of vegetable cultivars,
which it will soon be obligatory for seed nurserymen to use,
is contributing to the cultivar drain. Many older or little
grown cultivars have not been included in the lists and will
undoubtedly disappear quickly from cultivation unless efforts
are made to preserve them. If the newer cultivars were
improvements in all characters (including flavour in particular) no-one would worry about the disappearance of the
older cultivars – but in many instances this is not the
case.

The loss of cultivars is apparent in all aspects of horticulture whether one is dealing with ornamentals, fruit or
vegetables. With large-scale agricultural crops the effect
is perhaps of less significance as some of the problems
have been studied and in certain instances stocks of older
cultivars are purposely maintained. But with many ornamentals, even those of very considerable economic importance,
little or no thought has been given to the importance of
conservation.

If the need is accepted how far does one go in conserving
cultivated plants of particular genera?

It is certainly not my intention to suggest that every
cultivar should be preserved regardless of its value or
potential. That would be an impossible and pointless task
particularly when one looks at the multitude of names which
have been applied to such genera as <u>Narcissus</u>, <u>Rhododendron</u>
or <u>Rosa</u>!

4. Conservation of horticultural plants must be looked at
   practically and not emotionally. Each group has its specialists who should be able to draw up lists of cultivars
   of value historically, or in breeding, or for some other
   valid reason. In some instances International Registration
   Authorities (IRA's) exist for certain genera or groups and
   produce International Registration lists of cultivar names.
   Any Conservation policy could be linked without difficulty
   to these International Registers which could indicate
   "conserved cultivars" and provide information on the location
   of the living collections of the cultivars concerned.

The maintenance of such collections would undoubtedly
provide problems - but no more so than the maintenance of
threatened wild species.   Botanic gardens, arboreta and
some nurseries could co-operate on such a project and in
Britain the National Trust Gardens and perhaps Parks
departments could be involved.   In the case of annuals and
biennials where the cultivars are seed-raised a procedure
similar to that of the Kew Seed Bank could be adopted and
this would be of particular importance for certain
vegetables.

Such collections would involve considerable upkeep and the
provision of accurate documentation, including, where
practical, the preservation of properly annotated herbarium
specimens.   Most of us accept almost without question that
costly museum facilities should be provided for a very wide
range of objects of interest to men and it seems perfectly
reasonable to me to suggest that cultivated plants should
be accorded similar facilities by the community as their
importance economically and aesthetically is undeniable.

The case I have put forward is over-simplified and it has not
been possible to go into detail on the implementation of any
conservation policy for cultivated plants.

I am convinced, however, of the urgent need for such conservatio
which could be of considerable practical value in contributing to
Man's needs provided that it is not all-embracing and is carried
out realistically.

INTRODUCTORY PAPER II

G. S. Thomas

The National Trust

England

Species may be rare in nature or inaccessible due to countries
being closed to travel but cultivars become rare for a variety of
other reasons.   In addition, while taxonomic changes may result
in a variation of the quantity of species, they are in a way finite,
while the range of cultivars is infinite.   Thus I am pleased to
be able to make this contribution in connection with cultivars.

During the time spent on preparing this paper I have found
myself trying to assess the fundamentals which bring us all
together with the one idea of preservation.   The reasons are deep
and far-reaching and I should like to make a brief review of them.

The garden started as a space in and around which nourishing
and useful plants were grown.   If a community exhausted the local
supply of a valued herb, plants would be brought from farther
afield and cultivated near the dwellings.   This was undoubtedly
the first sort of conservation, and it was done by gardeners.
Plants which appealed to the senses would likewise have joined
necessary plants later on.   In our country we have evidence that
in mediaeval and Tudor times, when our gardening history started,
plants of all kinds were grown together in beds and borders in
hedged or fenced enclosures and later in walled gardens.   The
people who grew these plants were called gardeners and herbalists.
Gradually the herbalists and those who studied plants acquired the
term of botanists, and became scientists, while gardeners divided
in a similar gradual way into those who specialised in cultivation
and production and those who sought to use plants for the sake of
their effect.   The extremes are the ultra scientific botanists
and the plant artists of today, yet we are all still united by our
love for plants.

For thousands of years mankind has treasured forms of plants
which were to his ideas more useful or beautiful than the norm.
They may have been mutations, seedlings or chance hybrids.   Many
from very early days still survive, such as the olive, the orange,
the Madonna lily and German iris.   In England ancient cultivars
which illustrate the development of today's apples remain in
cultivation; France is more famous for its ancient pears and roses.
The Dutch yellow crocus originated some 250 years ago and is still
widely grown.

In addition to plants such as these which have developed within
our Western civilisation, a constant stream of species and cultivars
has reached our gardens from all temperate parts of the world.
It is these new plants, both species and cultivars that have given
a constant fillip to gardening through the centuries, for novelty
is always the spice in changes of fashion.   The last hundred years
have seen an unprecedented range of plants being brought into
cultivation in Europe alone, and this great stream of novelties is
not likely ever to be repeated, since most of the world's
temperate areas have been fairly thoroughly combed.

I think we may say that up to 1914 the craft of gardening in
England had been one long and gentle progression; many of the same
skills were practised as in Roman times.   An equally continuous
progression had occurred in selecting and later in hybridising
plants.   In spite of the radical changes in the fashion of garden
design from 1650 to 1830 or thereabouts in this country - veering
from extreme formality to extreme informality - all types of plants
were preserved in the great walled kitchen gardens, and also of
course in cottage and villa gardens, where fashion in design did
not intrude.   And so for a very long time the areas and skills
remained the same.

Since 1918 there has not simply been gradual change but an
acceleration in everything.   This change in horticulture has been
due to three quite different and major factors.   The first is the
rise in quantity, status and expertise of the owner gardeners:
they have not been trained in a tradition of gardening like the old
head gardeners, conservative to the core.   Rather are they open to
every whim and fancy that blows.   These whims and fancies have
been fostered by the next factor - the publicity given to gardening
by books, the press, radio and television, all of which thirst for
news, particularly in the shape of novelties.   The third factor is
the novelties themselves - derived by hybridisation, the only major
source of new plants left to us who garden out of doors.   Hybridisation and selection of better forms has had a great effect on the
yield from plants which were first grown for food; it was helpful
originally for subsistence.   Later it has pandered to greed for
bigger and brighter flowers regardless of nature's original form,
more prolific vegetables and bigger and better fruits.

Unfortunately breeding has usually proceeded along stereotyped lines, witness for instance dahlias and roses.   In 1906 a list was published of some 12,000 cultivars of roses - only they were called varieties; most of them have been lost and heaven knows how many have been raised and named since.   They come and go with alarming rapidity.   Nine tenths are not worth preserving, fortunately, because it is beyond our power to preserve everything.

The landmarks in hybridisation are extremely important both from a historical point of view and also when considering parentage in a scientific light.   First crosses with species outside the main-stream are therefore of considerable interest.   Likewise strains that have been taken up with enthusiasm and then dropped owing to a change in fashion, but which may regain favour: it is important to safe-guard a few of each.   Examples are to be found in un-frilled sweet peas, (Lathyrus odoratus), michaelmas daisies (Aster novi-belgi), montbretias (Crocosmia x crocosmiiflora), diploid bearded irises, violets double and single, purple, white, pink and yellow; the favourite ancestral European roses, and foxgloves (Digitalis) with flowers arranged in the normal style.

Private gardens have always played a big part in the preservation of rare plants, but often not deliberately.   The mere fact that there are thousands of different tastes, apart from climates and local conditions all over the world makes this obvious, and very choice and rare plants are usually acquired by exchange.   Every garden of any standing - and many a small one too - is a potential source of rare plants.   I visit many a garden in these islands and seldom do I go to a fresh one without seeing something new to me. In these gardens there is an immense reservoir of species and cultivars, untapped, unrecorded, even unknown.

Now I should like to mention the gardens of the National Trust and the National Trust for Scotland.   What I have already stated applies as much to these Trust gardens as to those still in private ownership, except that we in the Trusts do know what rare plants we have.   The National Trust is what may be called the largest private gardener in the world, and has not only gardens of every kind but a vast accumulation of plants gathered together by the original owners and successive generations.   The Trust seeks to foster collections started in separate gardens, enlarging upon them and even starting others where appropriate.   It has in addition to extensive collections of magnolias, rhododendrons, and camellias in many gardens, good collections of Acer, Sorbus, Yucca, ancient European roses, Hypericum, Deutzia, Hydrangea - some eight forms of Convallaria majalis and fifteen of Anemone nemorosa.   Besides separate genera, there are large assemblies of autumn colouring trees and shrubs, plants with 'Atropurpurea' foliage, etc.   The safe-guarding and elaboration of these collections is largely in the hands of the Garden Advisory staff, and we like to think there will always be such experts in the future.

For many years I have thought it would be highly desirable that there should be some sort of concerted national movement towards the conservation of garden plants in this country.   No single organisation could tackle it but a scheme whereby each botanic garden, horticultural institution, public park, National Trust garden etc, might take on one or more genera would I think be desirable.   They would collect the cultivars, sort and verify them, label and map them for all to see, in whichever part of the country was deemed most suitable for their growth and health.   I wrote a letter some years ago along these lines to the "Gardeners' Chronicle", but the response was negligible.

There is, however, cheering news.   The Garden History Society is compiling a list of scarce and important cultivars.   The National Trust is busy recording all the major plants in its gardens.   Even better is the recent offer by Kew to help in recording and computerising the lists.   The scheme might in time cover all major gardens, and would start with species of known provenance.   The National Trust is keen to cooperate and considers cultivars should next be tackled, rather than the enormous duplication in gardens of clones of species purchased over the years from nurseries.   There is at present the stumbling block of finance, which we hope may be overcome.   It would be a great endeavour and a goal worth achieving.   We are very fortunate to have in the files and archives of the Royal Horticultural Society records of origin and descriptive particulars of innumerable cultivars, submitted from time to time for trial at Wisley.

But lists and cards, maps and computers cannot do it all.   The whole scheme could founder for want of dedicated plantsmen. Therefore you and I have a duty to perform - to encourage the students of horticulture along the paths of plantsmanship. Fundamentally, in other words, it is necessary to bring home to the young that it is the beauty of plants which brings us all together, and this should be the reason for seeking a living from horticulture; all else in the vast and complex horticulture of today is subsidiary to it.

Therefore I give you five watchwords; we should each -

1. assemble our own speciality

2. record all items

3. preserve them

4. increase them

5. disperse them. Give them away. Apart from the delight of
   sharing beauty with others, it is the surest way of preserving plants. It also acts as an insurance against your own
   plants meeting with theft, ill health or disaster.

DISCUSSION SECTION V CONSERVATION OF HORTICULTURAL PLANTS

Mr Hillier (Winchester) mentioned the importance of commercial nurserymen in the introduction of exotic species of many plant groups, and listed some of the famous nurseries and their specialist introductions from overseas. He also pointed out the change in recent years caused by economic pressure, and the consequent production of massive numbers of increasingly fewer cultivars.

Dr Stearn (British Museum, Natural History) talking of his long association with botanic gardens, said that many plants to his knowledge had disappeared from cultivation. It was of no use bringing plants into gardens if the gardens could not grow them or keep them. He quoted examples of plants of no decorative value which were nevertheless of great importance in the history of genetics. They were the laciniate form of Chelidonium majus and the entire leaved form of Urtica pilulifera, two of the oldest mutations known and hence of great evolutionary and historical significance. It was the duty of those in charge of living collections to tell the gardeners themselves about the rare species so that they would have a vested interest in maintaining them.

He also spoke of the scientific role of botanic gardens such as in providing plants for biochemical study, mentioning his own work on the Apocynaceae and cancer. He pointed out that fewer European species of Allium were cultivated now than were grown in pre-war days, and said that it was wrong for a botanic garden to discard a plant without checking first whether it was the last specimen in cultivation.

Professor Raven (Missouri) spoke of the lack of co-ordination between gardens in the United States and suggested that this was due partly to the fact that many gardens are privately supported and do not have a large research component. He talked of the need to know what was held in each garden and stressed the importance of co-ordination on an international basis between botanic gardens.

Professor Zohary (Jerusalem) mentioned the possibilities of introducing horticultural plants to developing countries without a history of horticulture. Until now such introductions had been very haphazard. A better way of assessing the climatic regions of the countries concerned was needed, in order to know better which

plants would flourish. At present horticulture is confined to only
a few countries and there is a major possibility of conserving
plant material elsewhere.

Sir Otto Frankel (Canberra) spoke of his involvement in the
conservation of obscure cultivars of crop plants. Should they be
preserved? He felt it was only the germ of future evolution that
was really relevant to the future. For example he expressed a
wish to see the early sugar beets which are unknown but must have
resembled Beta maritima. The carrots of the Middle Ages are also
unknown. In contrast the plants of Philipe de Vilmorins, the
first breeder of pure lines, have been maintained and the
historical details are known. He felt less sure of the need to
maintain the two mutants referred to by Dr Stearn when there was
so much of vital importance to preserve for science and the future.
Concluding, he said that he would like to see the following
preserved: (1) plants of real historical value in the evolution
of horticultural plants; (2) those with known genetic value;
(3) those with known resistance to pests and diseases, very little
being known of the natural resistance of horticultural plants
because of the extensive use of chemicals to control pests and
diseases. He spoke of his support for conservation of species but
felt that it was not incumbent on us to extend the life of a
cultivated species indefinitely partly because different breeding
methods may be used in the future.

Mr Halliwell (Kew) suggested that the press, radio and
television should be encouraged to popularise unusual plants which
are either unknown or at present unpopular but which are worthy of
conservation.

Techniques of Collecting

SAMPLING GENE POOLS

J.G. Hawkes

University of Birmingham

England

The methodology of living plant collecting has changed very considerably in the last few decades. Not so long ago botanic gardens, arboreta and similar organisations considered that they were doing what was required of them if they grew a single accession of as many species as possible. Even if they possessed the facilities for conserving more than one collection of any particular taxon it was seldom thought necessary to grow more than two or three accessions of each species as a "representative" sample. Indeed, both the gardens and the collectors, basing themselves perhaps unconsciously on the practice and theory of classical taxonomy, assumed that each species or named variant of it was completely uniform. Hence for each uniform taxon only one collection would be needed. For example, with an alpine plant which possessed a polytopic distribution in the Alps, the Pyrenees, the Carpathians and perhaps some other outlying massifs, possibly a living collection from each mountain range might be considered to be of interest, especially if some distinct morphological differences could be discerned in each area which merited distinct varietal or form names. Ubiquitous species might well not be collected at all or at best might be represented by one collection only.

Biosystematic approaches to species studies have changed this outlook completely by demonstrating the enormous range of genetic variation incorporated in all but the very rarest species, and calling attention to the population concept in the assessment of species evolution. When morphological, cyto-genetical, biochemical and other features, in common and widespread species are carefully measured and analysed, they are found to vary between individuals in populations and perhaps even more throughout their range. Some variation is smooth and can be described in the form of clines,

which may well vary for different characters in different geographical directions.   Other variation patterns may show little
change in some parts and very abrupt and striking changes in other
parts of the distribution area.   Those of us who have worked with
domesticated plants such as wheat, rice, barley and potatoes, as
well as wild species related to them, have come to regard it as
axiomatic that it is extremely dangerous to base sweeping statements on the nature of such species from the results of analysing
just a few samples.   Disease resistance may be present in some
samples and absent in others, as has been shown in very many
instances with crop plants and related wild species.   Useful or
undesirable biochemical features may be present in high or low
amounts, or may be absent altogether, according to the samples
analysed.   The varying amounts of alkaloids in <u>Solanum dulcamara</u>
populations and the differences in cyanogenic glucosides in
<u>Trifolium</u> and <u>Lotus</u> species as well-known examples.

Sometimes, when enough samples are examined, significant and
valuable geographical patterns can be traced in a particular species
or in a group of related or even unrelated ones in certain areas,
as N.I. Vavilov (1926) has pointed out.

Such variation is not confined to herbaceous species.   Thus
forest research scientists have established great genetic diversity
in timber quality, adaptation to different environments and other
characters within the same tree species.

Even when there is no question of economically important
variation, detailed studies, as the classical works of Clausen,
Keck and Hiesey (1940, 1945, 1948) and so many other workers have
shown, will reveal morphological, biochemical, and ecophysiological
differences between populations in different parts of the range of
any species one cares to examine.

It should now be standard practice for field workers to conside
the need for at least some degree of multiple sampling, so as to
capture as large a part as possible of the genetic diversity within
species.   This is particularly necessary when species are being
destroyed or at least are under threat.   Much research directed
towards the understanding of evolutionary patterns, ecological
adaptation and biochemical variation is needed in most species;
furthermore, many may be of potential economic value.   Wide
sampling of gene pools will help to preserve this variability, then,
for future studies.

Multiple gene pool sampling is particularly necessary at
present, and indeed has been so for a number of years, since even
quite common species, and certainly the rarer ones, though perhaps
not always under immediate threat of extinction, are nevertheless
suffering what has been described as "genetic erosion".   By this

is meant that the full original range of genetic variability has
been diminished by destruction of vast sections of the distribution
areas of these species, which have thus suffered erosion of these
sections of their forms of genetic diversity.   As forests are
felled, marshes are drained, sea coasts are turned into holiday
resorts, mountain pastures are trampled and grazed, heathlands are
changed into grassland, old rich meadows are ploughed up and planted
to crops or re-seeded with standard grass mixtures, cattle and
sheep grazing is intensified, cities expand, industry spreads and
roads are widened, so the total genetic diversity of all plants in
every part of the world is diminished.

Other speakers in this symposium are dealing with conservation
as such.   One of the most important aspects of this is to ensure
that what is left now of the genetic diversity of wild and cultivated species is sampled in such a way as to preserve as much as is
reasonably possible of this diversity for the future.

It may well be objected that too intensive a programme of
sampling will defeat the very ends we seek, and will hasten the
destruction of wild species.   The ethical aspects of this are
considered in the paper by Raven but the undermentioned points may
be considered in the context of this paper.

If a species is on the way to extinction and perhaps can only
be ultimately preserved in a limited number of special reserves,
judicial sampling within the highly threatened parts of its range
may serve an extremely useful function, in helping to preserve a
greater measure of variability than would otherwise be possible.
Where only a few scattered stands of an extremely rare species are
known, sampling by seed collecting will act as an insurance against
the complete destruction in the field of these last remaining plants,
since possible ecosystem reconstruction may later be possible, using
the seed bank material as described in the paper by Professor
Hondelman.   Of course this is not so necessary if adequate biosphere
reserves are established.

Methods of sampling depend on the species concerned, its
reproductive biology and other factors.   In general, it is much
better to collect seeds, since they represent a much wider sample of
the gene pool, weight for weight, than other parts of the plant.
They are easier to collect, providing that the collector judges
accurately the timing of his collecting expedition, and once
collected they can be transported to their permanent store with
comparatively little risk of damage.   Furthermore, in a normally
cross-pollinating species the seeds from one plant will represent a
wider sample than the vegetative organs, which from any one plant
will of course be genetically completely uniform.

With species reproducing largely vegetatively, such as many

members of the families <u>Liliaceae</u>, <u>Iridaceae</u> and <u>Amaryllidaceae</u>, much collecting of bulbs, corms and rhizomes will be necessary. Nevertheless, when seeds are available, it is strongly recommended that these should be collected, rather than the vegetative organs.

The reasons for this suggestion lie not only in the already mentioned need to sample the local gene pools as adequately as possible.   Transport and quarantine problems also make vegetative sampling difficult, whilst the digging up of samples at a time when they are in a dormant state means that a previous visit to identify the plants when they are in flower would ideally be necessary. The slowness with which underground storage organs can be obtained means of course that fewer samples will be made, thus reducing the likelihood of adequate gene pool sampling.   Yet the sampling of vegetative material is often of great scientific and practical importance.   Chromosome races and rather infertile aneuploids resulting from past hybridisation may be gathered more easily and will show a richer range of variation from vegetative collections. Specially adapted genotypes in largely clonally propagated material would be collected by taking vegetative organs but not be seed gathering.   In a poor year for seed and fruit setting the collector will be forced to make vegetative collections; and where seeds become quickly inviable, as we shall mention below, again, vegetative organs are the only possibility.

Unfortunately, not all plants possess seeds which remain viable when removed from the plant and cooled and dried under the usual conditions.   Seeds which die quickly under normal treatment are termed "recalcitrant" and need special methods of storage which are still imperfectly understood.   Temperate fruit and nut trees, most tropical fruits, many forest trees and plants of considerable economic value such as coffee and cocoa have shortlived seeds of this type.   At present, the only solution is to make cuttings for grafting or rooting, or to dig up seedlings if any can be found.   The problems of transport and maintenance are clearly enormous with this type of material but they must be faced, and research inputs are urgently needed in this area to devise the best methods of long-term cultivation or of storage in the vegetative state.

We should now consider in some detail the problem of sampling. We have already agreed that multiple sampling is imperative if a reasonable spectrum of the genetic variability is to be captured. How is this to be done?

In a recent paper, Marshall and Brown (1975) consider the problem of optimum sampling strategies in order to collect adequately the genetic variation within species.   Defining as their aim the collection of at least one copy of each variant with a frequency in the population greater than 5% they advocate the following: A bulk sample of seeds from about fifty plants should be taken as a unit

collection from each site.   Several of such site samples should
be made in a clustered manner, rather than evenly, through the
distribution area of the species.   Where the clusters are taken
the collector would depend on his judgement and would try to collect
material from as many different habitats as possible.   If the
habitats are very varied then more samples should be taken; if the
habitat is uniform, fewer will probably suffice.   If the breeding
system is known, fewer samples of outbreeding species will be
needed in each area, and more of inbreeding ones.   Unfortunately,
however, in many, perhaps most, cases this information will not be
available.

   To sum up then, the plant explorer should endeavour wherever
possible to take bulked seed samples from up to about fifty
individuals, and certainly no more than one hundred.   Samples
should be clustered and the clusters should be spread as evenly as
possible throughout the collecting area.

   Of course, all this advice may well be regarded by the practical
collector as a "council of perfection".   Nevertheless, it serves to
show what should ideally be the collector's aim, and the collector
should certainly try to concentrate on areas where habitat differences such as soil pH, moisture, exposure, climate and altitude
differences are great and to make fewer collections where the
habitat seems to be comparatively uniform.   Bulked seed samples from
up to fifty plants should provide adequate sampling of the local
gene pools.

   Where concentration of effort needs to be directed towards the
collection of a few species of economic or scientific value Jain
(1975) recommends, if possible, at least two collecting trips to
the same area.   In the first a "coarse grid" series of collections
is made at wide intervals over the whole area.   The variation in
samples from this is later studied in the experimental field.   In
collections from areas where interesting or intensive variation
appears, then the second expedition would provide "fine grid"
collecting by concentrating the sampling in these regions.   Perhaps
this also may be regarded as impossibly complex and time-consuming
by the average collector.   I can only reiterate that it is a method
that should be attempted wherever possible, so that the better the
attainment, the more chance there is of adequate gene pool sampling.

   Finally, we must consider the problem of field records.   Here
again, as with sampling, the greater the concentration of effort and
time on one area or sample, the less ground will be covered.   However, whereas the sampling is of positive value in bringing more
variability into the collections, the recording of data, if very
time-consuming, can be very counterproductive.   A number of complex
recording systems have been advocated, (see for example, chapters
by Bunting and Kuckuck and by Godron and Poissonet in Frankel and

Bennett, 1970 and by Konzak in Hawkes and Lange, 1973, pp 28-36).
The average collector would find such systems somewhat over time-
consuming, and in the field much information would simply not be
entered.   To put it crudely perhaps, one has to make a choice
between collecting plants or collecting information!

At the other extreme, collectors in the past, though rarely
nowadays, have merely put the name of the country or even sub-
continent on their seed packets or herbarium specimens.   Clearly,
a useful compromise must be sought, so that most of the available
collecting time can be spent in gathering plants rather than filling
up elaborate forms asking for complex topographical and soil information, such as details of slope, aspect, stoniness, physical texture,
colour, depth, pH etc., and climatic data such as temperature,
rainfall etc., most of which is impossible to be certain of during
an expedition in any case.

On the other hand a basic minimum of information is not only
useful but indeed essential.   Printed notebooks are valuable in
helping to remind one of what information is needed but they should
be as simple as possible.   No field collector under extremes of
heat, humidity, stinging insects or blood-sucking leeches has the
time or inclination to write a short monograph on each plant he
collects.

The following data may, however, be considered basically
essential (Figs. 1 and 2):

1.   Name of collector and collecting number

2.   Latin name of species (or genus or family at least)

3.   Vernacular name

4.   Provenance data, e.g. Country, province, district, nearest
     town or village and distance and approximate direction from
     it (e.g. 10 km from a certain town in a S.E. direction on
     trail or road to another town, by a certain river or on a
     mountain range).

5.   Latitude and longitude (to be filled in later from a map)

6.   Date of collection

7.   Altitude of collection

8.   Photo number

9.   Type of material collected, e.g. seeds, tubers, cuttings,
     plants, herbarium, etc.   The need for voucher herbarium
     specimen wherever possible should be stressed.

**INSTITUTION
OR
ORGANIZATION**

Collector: .... No. ....

Latin name: ....

Vernacular name: ....

Locality: ....

....

....

Latitude: ....° ....ʹ   Longtitude: ....° ....ʹ

Date: .... Altitude: ....m   Photo No: ....

Material: Seed   Vegetative parts   Herbarium   Plant

Status: Wild   Weed   Cultivated

Frequency: Abundant   Frequent   Occasional   Rare

Habitat: ....

....

Description: ....

....

Disease Symptoms: ....

....

Other Data: ....

Team: ....

Collector	Collector	Collector	Collector
No:	No:	No:	No:

Figure 1

10. Status, e.g. wild, weed, cultivated, escaped, etc.

11. Frequency – a rough estimate is probably all that is
    required, e.g. abundant, frequent, occasional, rare.

12. Habitat, set out in simple terms such as: hedgerow,
    swamp forest, savanna, etc.

13. Special descriptive features that cannot be preserved in
    the herbarium specimens of seeds collected, such as height
    of tree, type of branching, colour, size and shape of
    fruits, flower colour and shape, leaf and stem colour,
    leaf shape if too large to fit on to a standard herbarium
    sheet, should be noted but not details which can be seen
    on the specimens collected.

14. Any other features of interest, if known, for example
    medicinal or economic properties, supposed disease,
    drought, cold, etc. resistance.

Finally, it may not be thought necessary to point out that a
single consecutive numbering system should be used. Even so, other
complex and confusing systems have been proposed which are very
unsatisfactory and in my opinion should be avoided at all costs.

A field recording sheet is shown in Fig. 1 and a completed
recording sheet in Fig. 2. These can be printed and bound in sets
of one hundred into notebooks with tear-out labels to put with the
collections.

To sum up, then, the plant explorer who wishes to collect living
material should sample populations rather than single individuals,
and should make multiple samples from as much as possible of the
ranges of the species in which he is interested. Seed collections
should be taken wherever possible, though with "recalcitrant" or
short-lived seeds other methods of collecting may be necessary.

Finally, adequate field data should be taken and noted on pre-
printed pages with pre-stamped numbers in notebooks of convenient
size with tear-out labels. Botanists will differ on how much or
how little data should be taken, but a certain minimum of information, as discussed here, would seem to be essential in order that
further material may be collected by the same or other explorers
and that general conclusions as to distribution, altitude and flowering time frequency and habitat preference may be reached when the
material is described in research papers and monographs or used by
breeders and economic botanists.

NETHERLANDS/BIRMINGHAM UNIVERSITY
POTATO COLLECTING EXPEDITION, 1974.

———

Collector: Hawkes, van Harten + Landeo   *No.* 5500

Latin name: S. acaule.

Vernacular name:

Locality: Dept. Puno, prov. Huancané, 11 km.de Huancané en el camino hacia Moho y ± 2km después de Vilque Chicc.

Latitude: 15 ° 13 '   *Longtitude:* 69 ° 41

Date: 14/3   *Altitude:* 3820 m   *Photo No:*

Material: (Seed)   Vegetative parts   (Herbarium)   Plant

Status: (Wild)   Weed   Cultivated

Frequency: (Abundant)   Frequent   Occasional   Rare

Habitat: Stony places in open ground and under bushes (Muña, etc).

Description: Single plant collection made to test the possible inbreeding mechanism in S. acaule.

Disease Symptoms: None.

Other Data:

Team:

Collector	Collector	Collector
No: 5500	*No:* 5500	*No:*

Figure 2

REFERENCES

CLAUSEN, J., KECK, D.D. and HIESEY, W.M. (1940). Experimental studies on the nature of species. I. The effect of varied environment on western N. American plants. Publications of the Carnegie Institution of Washington 520: 1-452.

CLAUSEN, J., KECK, D.D. and HIESEY, W.M. (1945). Experimental studies on the nature of species. II. Plant evolution through amphi-diploidy and autoploidy with examples from the Madiinae. Publications of the Carnegie Institution of Washington 564: 1-174.

CLAUSEN, J., KECK, D.D. and HIESEY, W.M. (1948). Experimental studies on the nature of species. III. Environmental responses of climatic races of Achillea. Publications of the Carnegie Institution of Washington 581: 1-129.

FRANKEL, O.H. and BENNETT, E. (eds.) (1970). Genetic resources in plants - their exploration and conservation. I.B.P. Handbook No. 11. Blackwell Scientific Publications, Oxford.

HAWKES, J.G. and LANGE, W. (eds.) (1973). European and regional gene banks: Proceedings of a Conference of Eucarpia, Izmir, Turkey 1972. Wageningen, Netherlands.

JAIN, S.K. (1975). Population structure and the effects of breeding system. In FRANKEL, O.H. and HAWKES, J.G. (eds.). Crop genetic resources for today and tomorrow: 15-36. I.B.P. 2, Cambridge University Press.

MARSHALL, D.R. and BROWN, A.H.D. (1975). Optimum sampling strategies in genetic conservation. In FRANKEL, O.H. and HAWKES, J.G. (eds.). Crop genetic resources for today and tomorrow: 53-80. I.B.P. 2, Cambridge University Press.

VAVILOV, N.I. (1926). Studies on the origin of cultivated plants. Bulletin of Applied Botany (U.S.S.R.) 16(2): 1-248.

ETHICS AND ATTITUDES

Peter H. Raven

Missouri Botanical Garden

United States of America

ACKNOWLEDGEMENTS

I am most grateful to E.S. Ayensu, B. Bartholomew, L. Benson, D. Bramwell, G. Davidse, P.H. Davis, B. de Winter, J.M. Diamond, C.H. Dodson, R.L. Dressler, Hj. Eichler, J. Elsley, F.R. Fosberg, A. Gentry, G.W. Gillett, D.R. Given, P. Goldblatt, L.D. Gottlieb, W. Greuter, A.V. Hall, L.R. Heckard, N. Hepper, R.A. Howard, H.W. Irwin, W. Jankowitz, L.A.S. Johnson, M.C. Johnston, G.C. Kennedy, L.W. Lenz, W. Meijer, R.V. Moran, O. Nilsson, M. Ospina, G.F.S. Pabst, F. Perring, J. Rourke, H.B. Rycroft, A.K. Skvortsov, W.H. Wagner, Jr, J.S. Womersley, K.R. Woolliams and P.R. Wycherley, for the valuable advice they have given me in the preparation of this paper.

The roughly 300,000 species of green plants and algae provide the means by which the energy of the sun that reaches the earth's surface is locked up in chemical bonds. By carrying out this process, the plants and algae provide all of the food for from ten to thirty times as many heterotrophic organisms, including all the animals and man himself. Because of the nature of food chains and the degree of specialization in feeding habits of these heterotrophs, we may assume that the extinction of each species of plant is, on the average, accompanied by a ten- to thirty-fold loss amongst other organisms. Therefore, the diversity of plants is the underlying factor controlling the diversity of other organisms and thus the stability of the world ecosystem. On these grounds alone, the conservation of the plant world is ultimately a matter of survival for the human race.

     As stressed elsewhere in these proceedings by
Dr Gerardo Budowski, Director General of the International Union
for the Conservation of Nature and Natural Resources, living
collections of plants can be invaluable for conservation purposes
if their management is based on a sense of purpose backed by
scientific data and appropriate techniques. The importance of the
role of botanical gardens with respect to this problem is under-
scored by a resolution passed by the International Association of
Botanical Gardens meeting at Moscow in July 1975. Before we can
confidently embark upon the construction of a global network of
institutions for this purpose, however, many thorny questions must
be answered. How many species do we wish to preserve, and for what
purpose? What standards of collecting and growing them are
acceptable? What is the social and economic justification for such
preservation? And how can we stress the preservation of plants in
botanical gardens without correspondingly decreasing our enthusiasm
for preserving them in the wild? As a basis for attempting to
provide answers to these questions, we must consider a number of
pertinent facts.

     While preparing this paper, I have consulted colleagues
throughout the world, and it has become evident to me that
differences exist between different regions, and that these must
profoundly affect our conclusions. In the following pages, I shall
first briefly review the scope of the problem, with emphasis on
these regional differences; then discuss some aspects of the
transfer of wild plants to botanical gardens and their treatment
once they are there; and conclude with a brief review of the
current status of commercial exploitation of wild plants.

THE SCOPE OF THE PROBLEM:

The Tropics

     Taking a relatively conservative view and assuming that there
are about 240,000 species of flowering plants in the world, we may
very roughly estimate that something on the order of 50,000 species
inhabit the North Temperate zone, including northern and central
Mexico, North Africa, and the Himalayas; at least 15,000 temperate
to arid Australia and New Zealand; roughly 10,000 the Cape Region
of South Africa and bordering arid zones; and about 10,000
temperate South America. The remaining 150,000 species are tropical
in the broad sense, with perhaps 30,000 in the tropical and semi-
arid regions of Africa, including Madagascar; at least 35,000 in
tropical Asia, including New Guinea and tropical Australia; and
the staggering figure of perhaps 90,000 species in South America,
Central America and tropical North America. Thus nearly two-thirds
of the world's plant species appear to be tropical in distribution.

How many of these species can be considered threatened at present? For the temperate regions of the world in general and the bordering arid zones, it appears likely that something of the order of 4,500 species could strictly be considered threatened out of the total of about 85,000 for the temperate regions of the entire world. About a third of these appear to be in the Cape Region of South Africa, where they comprise at least a quarter of the entire flora. There are perhaps 750 of these 4,500 species which should be considered endangered or vulnerable in the strict sense of the IUCN Threatened Plants Committee (TPC). (These statements are calculated from Walters, 1971; Specht, Roe and Boughton, 1974; Anonymous, 1975a; Perring, 1975; and personal communications from A.V. Hall, G.L. Lucas and H. Synge). The usage of these terms within the United States (Anonymous, 1975a) is certainly broader than that now advocated by the TPC, and would suggest a total of perhaps 6,000 threatened and 1,500 endangered and vulnerable species for the world's temperate regions, but the order of magnitude is the same.

For the approximately 155,000 species of flowering plants that are tropical in distribution, the problem is entirely different, and is one of the widespread destruction of habitats coupled with a grossly inadequate knowledge of the plants. In addition, botanical gardens are much less numerous in the tropics than elsewhere, and in general much more poorly funded than those in the temperate zone. These are alarming facts if one also takes into account that the populations of Latin America (currently 325 million), tropical Asia (1.2 thousand million), and tropical Africa (275 million) will double by the end of the present century, contributing about eighty-three per cent of a world population increase from its present level of four thousand million to an anticipated 6.25 thousand million in the year 2000 (1975 World Population Data Sheet, Population Reference Bureau, Inc.). We are just beginning to chart the principles upon which tropical ecology should be based (Farnworth and Golley, 1974), and these burgeoning population pressures threaten to overwhelm most tropical ecosystems before we even begin to understand them.

Considering that there will be no undisturbed tropical lowland forest anywhere in the world within twenty-five years, except for relatively small reserves, and considering that at least 150,000 species of flowering plants are tropical in distribution, it is certain that many of them will become extinct during our lifetime (Anonymous, 1974). To illustrate this point we need only mention that the FAO estimates that about ten million hectares of tropical forest are being felled annually. Added to this, we are abysmally ignorant of the present status of these plants: which particular species of New Guinea, or the Amazon Basin, are threatened, and what are their ranges? What knowledge we have available tells us very little about the extent of narrow endemism in the tropics.

There appears to be little basis for taking particular tropical
species into cultivation and attempting to preserve them, and no
basis in my opinion for estimating the numbers of threatened
plant species in the tropical areas of the world. One supposes,
however, that at least a third of them will be threatened or
extinct by the end of the century, with the first major wave of
extinction in South-east Asia and New Guinea within the next
decade.

Tropical plant species certainly cannot be maintained on this
scale in greenhouses outside of the tropics. Greenhouses are
costly structures which consume large amounts of increasingly
expensive energy and in which the plants are still subject to
freezing or baking in case of power failure or fuel shortage.
Some of the threatened tropical plants can be brought into cultivation in time in the relatively few and poorly funded botanical
gardens of the tropics, especially if this effort is aided internationally, but very many of these species are likely to be lost in
any case.

In theory, the establishment of large reserves in various
parts of the tropics would appear to be the most promising way to
achieve the conservation of plant and animal species. There are,
however, formidable obstacles that lie in the way. For example,
the FAO has estimated in a 1974 report that by 1985 at least
twenty-six countries, all tropical, with a total population of
365 million will be unable to supply on a per capita basis
sufficient food for their inhabitants to avoid gradual starvation.
How much of the resources of such countries will their governments
be able to divert to the creation of and subsequent protection of
sizeable nature reserves? It is important to attempt to find the
answer to this question in relation to the potential activities of
botanical gardens. Although some preliminary studies have been
conducted of the ways in which developing countries might be
compensated for maintaining environmental quality (eg. Nicholls,
1973), the will of nations such as the United States, Japan, and
the members of the EEC to provide support for the maintenance of
tropical nature reserves and botanical gardens has yet to be
demonstrated, while the activities of their large firms in
rapaciously exploiting the forests and other natural resources of
the tropics continues unabated. Even when there is a well conceived and comprehensive system of nature reserves, as in
Madagascar, they nevertheless are likely to be threatened eventually
by growing population pressures unless they are perceived as
relevant to the local population (Bernardi, 1974). It may be
relatively easy to establish natural reserves in developing
tropical countries, but it will certainly be quite another matter
to maintain them through the years.

Even if reserves are set up and maintained, there is a growing body of information which suggests that they will enjoy only limited success over the long term. As pointed out by Gómez-Pompa, Vázques-Yanes, and Guevara (1972), those small reserves and forest remainders that may persist in the tropics will not regenerate in a manner comparable with temperate ecosystems. Instead, their reduction to small reservoirs in a fluctuating environment is tantamount to total destruction. Tropical plants appear to be overwhelmingly outbreeders, and widely scattered, so that their reduction to small populations may not provide a sufficient population to allow them to survive (Elton, 1975). Furthermore, the newly developed theory of island biogeography, much of it directly based upon data gathered in the tropics, provides ample evidence that only reserves of considerable size are adequate to ensure the preservation of comprehensive samples of biological diversity for long periods of time (eg. Diamond, 1972, 1975; Terborgh, 1974, 1975; May, 1975), thus rendering the practical problem still more difficult. Even the division of a reserve by a highway may serve to lower considerably the number of species that can ultimately be sustained in it. Correspondingly, the reduction in size of any natural area for any purpose will inexorably lead to the loss of many species even if they are all present initially in carefully conceived reserves, and these reserves are managed in an ideal manner indefinitely. For these reasons we know that we shall certainly not be able to preserve many, probably most, species of plants and animals, for more than another century, and that we must choose and set priorities if we are to do an effective job for any of them.

The difficulty of making reasonable choices about areas suitable for preservation as reserves in the tropics are enormous, because we have so little information. For example, the most important large timber tree of the Atlantic slope of Nicaragua, which has also been collected once in Belize and once in Panama, has never been collected in flower but seems to represent an undescribed genus of Flacourtiaceae (L.O. Williams, pers. comm.). If logging operations continue, it will be extinct, at least in Nicaragua, within five years. A new and previously uncollected genus of birds was discovered on the island of Maui in the Hawaiian islands in the summer of 1973 (Casey and Jacobi, 1974). Perhaps most striking of all, a third living species of peccary (family Tayassuidae), representing a genus that had been known only from the Pleistocene, has just been discovered in the Gran Chocó of western Paraguay (Wetzel et al., 1975), where it is one of the major mammalian species hunted for food. With the all-weather surfacing of the Trans-Chocó Highway, now in progress, it is likely that neither the newly discovered peccary or many of the other interesting mammals of the Gran Chocó will survive for long. Despite this general lack of information, however, there are certain tropical regions, such as Mt. Kinabalu in Borneo, which are

obviously highly desirable for reserves. Even for the vast stretches
of the Amazon Basin, information about centres of endemism and other
critical areas is beginning to emerge (eg. Haffer, 1969; Vanzolini,
1970; Prance, 1973).

    In the face of these problems, I submit that it is highly
unlikely that botanical gardens will be able to make a substantial
direct contribution to the preservation of any but a very few
selected species of tropical plants by cultivating them. Those
which are of known scientific, economic, or horticultural interest
will doubtless take first priority and far more such plants await
detection in the tropics than elsewhere. Aside from this, botanical
gardens must certainly play a responsible role in channelling public
opinion toward the creation of suitable reserves in the tropics,
and toward the protection of those reserves once they are created.
In order that this be done successfully, it will be necessary to
consider the whole problem of tropical land-use on a wider basis
than has hitherto been done (Anonymous, 1975b).   Examples of
creative thinking about the massive problems involved have been
provided by Janzen (1973), who stresses the need to develop what
he calls sustained-yield tropical agroecosystems, and the vital
role of education in the public acceptance of ecological principles;
and by Budowski (1974), who urges much more careful study as a
prelude to the establishment of tree plantations in the tropics,
and then the use of abandoned, deforested lands for their creation.
The usage of modified tropical vegetation for many purposes appears
to offer the best hope for their survival in the developing
countries (cf. Tinker, 1974).

    We live in a time when many of the earth's species are being
destroyed, with permanent consequences to the future course of
evolution, as pointed out by Paul Richards nearly twenty-five years
ago. Realistically, we shall have no opportunity to determine with
any degree of precision which plants are rare and endangered in
most of the tropics before the whole ecosystems in which they exist
become depleted or even cease to exist in response to this global
disaster.

    It is certainly appropriate for tropical botanical gardens to
attempt to preserve properly documented living collections of the
vegetation of their areas, but they perhaps have an even more
important role to play in public education, in assisting in the
creation of suitable nature reserves, and in helping in the development of ecologically sound modes of land use in which samples of
organic diversity as comprehensive as possible may be preserved.
Conservation efforts should be centred in reserves and potential
reserve areas, and the removal of plants for cultivation - cultivation of any kind - from tropical areas that are being destroyed
probably should be actively promoted, inasmuch as it may offer some
hope for the survival of these plants. Indeed, the pressures of

horticultural collecting appear insignificant in the face of the destruction of ten million hectares of forest annually, and it would appear to be realistic to encourage such collecting everywhere in the tropics, except in actual and potential reserves.

In building up their living collections, tropical botanical gardens should not necessarily follow the models established centuries ago in temperate regions, particularly in the face of the crisis of purpose facing such gardens everywhere. They should look for new ways to make their gardens relevant and interesting to the people of their own countries and localities, who must both support and learn from these gardens. In particular, all botanical gardens should consider explicitly their role in the conservation of nature; we no longer have centuries in which to develop an awareness of man's ultimate and complete dependence on green plants.

The relatively well endowed botanical institutions of temperate regions might do best, in their efforts in the tropics, to continue to stress survey work, together with the development of information-retrieval systems adequate to cope with the masses of data generated by studies of this kind. It is of the utmost importance that foreign institutions conducting studies of any kind in the tropics do everything in their power to strengthen the educational and research capabilities of their sister institutions, for only in this way can results of truly long term significance be achieved (Anonymous, 1970; Budowski, 1974). They must also do whatever may be possible to encourage the conservation of nature in the countries where they are collecting plant specimens, as suggested by Budowski (this symposium).

Temperate Regions

In the remainder of this paper I shall deal principally with the plants of the temperate regions of the world and the bordering areas which they control. In these regions, with some 85,000 species of which perhaps 4,500 may be considered threatened by the application of strict criteria, a consensus about how to deal with the preservation of threatened species of plants and animals is beginning to emerge. Similar concerns are evident in the United States (Anonymous, 1975a); in the United Kingdom (Perring, 1975); in Australia (P.R. Wycherley, pers. comm.); in South Africa (A.V. Hall, pers. comm.); in Sweden, where Project Linné has been under way since 1972; and in the Soviet Union. It is worthwhile calling attention to some of these points of agreement before specifically addressing the question of the role of botanical gardens in the preservation of threatened plants.

First, there is a great need for public education about rare and endangered plant species. The conservation-orientated

activities of botanical gardens, like all of their other activities,
ultimately must rest on public support, and this in turn can come
only from a public knowledgeable about the problems involved. The
establishment and maintenance of natural or semi-natural areas by
botanical gardens can be of special importance in increasing public
awareness of wild plants and of nature in general, as should be
evident, for example, to anyone visiting South Africa's magnificent
Kirstenbosch Botanic Garden. In addition, wild plants whether
threatened or not, may be more easily able to become established
and to reproduce themselves in such situations, thus keeping both
the cost and the risk of losing them at a minimum.

        Second, research into the proper methods of cultivation of
wild plants, many of which are notoriously difficult to establish
in gardens and to maintain, is of great importance. This has been
brought out very clearly by Mr Woolliams (this symposium) for the
remarkable flora of the Hawaiian islands. When the last few
survivors of a particular species are identified, it is usually
much too late for such research. Threatened species in general
should be brought into cultivation by means of seeds and cuttings,
thus ensuring a broader sample of the diversity present. Plants
of such species should be transplanted from the wild to gardens
only when these plants are in imminent danger of destruction in the
wild. Botanical gardens must not become major consumers of the
very plants they are trying to protect. The results of research
on the cultivation of wild plants should be made widely available,
perhaps in part by means of seed lists, as suggested by
Professor Heywood (this symposium), although it may not necessarily
require formal publication.

        Third, inventories of the plants of the respective areas must
be undertaken before rational consideration can be given to their
preservation. Botanical gardens can play a vital role in organizing
a network of information about endangered plants, and particularly
endangered plants in cultivation (Heslop-Harrison, 1973). Within
this network should be stored information about the simple parameters of their reproductive physiology, as Professor Heslop-
Harrison has suggested elsewhere in this symposium. The excellent
scheme of Dr M. Hoffmann, also reported in these proceedings,
provides an example of what can be accomplished in relatively short
order and at reasonable cost with modern technology. At the
Missouri Botanical Garden, for example, with Dr Hoffmann's help,
it was possible to determine that 104 of the taxa in our relatively
small collection of orchids were unrepresented elsewhere in cultivation, a statistic that will certainly inspire us to pay special
attention to these plants and to attempt to distribute them to
other centres so as to better ensure their survival. Computerization
of the records, then, appears to be a logical step toward international recognition of a particular garden's activities, something
that may be important for local morale as well as for funding. The

Plant Records Centre (PRC) of the American Horticultural Society already includes some 135,000 records of plants in cultivation at a number of gardens, and might easily provide the framework on which an international system could be constructed.

Fourth, there is general agreement that the preservation of plants in their native habitat is greatly preferable to their preservation in cultivation. Once a given plant species has been brought into cultivation, it will begin to change under a whole new set of selective forces and in a direction that cannot initially be predicted; its genetic variability will automatically begin to decline. Continuing evolution and adaptation to natural conditions which provides the only real hope for long term survival, is feasible only within natural communities (Frankel, 1974). Even the first sowing of seeds in a garden is likely to yield an aggregation of individuals very different from that which is found in nature, owing to the very different selective pressures. In addition, a species is far more likely to survive when it is maintaining itself in natural surroundings than in cultivation, particularly since the degree of attention and particular emphases of the cultivators often vary so markedly through the years. Nevertheless, it is probably worth stating categorically at this point that it is better to have a few genetically depleted individuals of a given species remaining in cultivation than none at all.

Moving on now to general considerations, the existence of a threatened species on public lands should be noted and brought to the attention of the appropriate authorities. If it grows on private lands, the owners should be notified and the lands if possible acquired as small reserves or added to existing reserves that may already be in the public domain. In any event, the co-operation of the private landowners in protecting populations of rare and endangered species on their lands should certainly be sought (Wallace, 1975). When it occurs on public lands, steps should be taken to ensure the long range survival of the endangered species; these steps may often involve deliberate modification of the habitat in order to keep it suitable for the plants in question. Perring (1975) has pointed out that, even though Wicken Fen can claim to be the oldest nature reserve in Britain, both Liparis loeselii (L.) Rich. and Utricularia minor L. have become extinct there during the time it has been a reserve, because the importance of peat digging to maintain certain open habitats was not appreciated soon enough. In an analogous case, Teucrium scordium L. was discovered at Whittlesey in Cambridge in 1968, whereupon the area was declared a reserve. Wildfowlers who formerly tramped through the area, thus providing an open habitat for the species, were excluded, whereupon the Teucrium promptly went extinct. Fortunately, it had been brought into cultivation at the Cambridge Botanic Garden and will now be reintroduced, with

suitable measures taken to provide appropriate conditions for growth
As time passes, natural hybridisation may even cause certain species
to "disappear" on the reserves that were created for them: Ratcliffe
(1973) reports that Erica ciliaris L. and Dryopteris cristata (L.)
A. Gray in Britain might actually be decreasing for this reason.
This suggests the possible desirability of the elimination of
certain relatives of the threatened taxa, especially if these
related taxa have been introduced into the reserved area.

With respect to the size of reserves, the same sorts of
consideration that have been mentioned in our discussion of the
tropics are pertinent, and the same literature applies. In
regions of mediterranean climate, however, and on islands, where
there are often rich concentrations of narrowly restricted species
in sharply distinct habitats within a small area, these habitats
can sometimes be preserved in relatively small reserves. As we
have seen in the example of Wicken Fen, however, a small reserve
surrounded by agriculture land may be radically affected by water
table changes induced through irrigation or drainage, and the
reserve also may be especially liable to exposure to drift of
fertilizers, herbicides or pesticides, rendering the survival of
the endangered species doubtful. In consequence of similar
pressures, the survival of the Australian Cephalotus follicularis
Labill., the only member of its family, may be possible only in
botanical gardens – the bogs where it grows are being drained for
agricultural purposes throughout its range (Hj. Eichler, pers. comm.
In order to maintain many species in island reserves, the elimination of domestic or feral animals may be necessary, whereas in
other circumstances, grazing may be required for the preservation
of the species of which the reserve was established. Certainly,
the management of populations of threatened plants in reserves can
be successful. For example, in New Zealand the only population of
Ranunculus crithmifolius Hook.f. subsp. paucifolius (T. Kirk)
F.J.F. Fisher declined to as few as twenty plants in about 1940,
mostly by overgrazing, but was restored by the erection of a
stock-proof fence and by careful gathering and sowing of the
naturally produced seed until it numbers several hundred individuals
(Fisher, 1965).

The monitoring of threatened species must continue actively
after plans for their preservation have been put into effect. Some
of these species may continue to decline, while others will become
more frequent, and perhaps eventually be considered out of danger.
Once the decision has been taken to attempt the preservation of a
particular species in nature, the consequences of this decision
must be sustained indefinitely if the species is to continue to
survive. Botanical gardens should play a leading role in proposing,
helping to establish, and maintaining a network of reserves devoted
in part to the preservation of threatened plant species, as we shall
discuss further below.

Many are discussing the role of botanical gardens as co-ordinating institutions, with satellite reserves or gardens in which threatened plants can be preserved either as natives or in habitats similar to those in which they naturally occur. Such a system is under particularly active development by the National Botanic Gardens of South Africa and appears to offer great promise. Amateur horticulturists might be enlisted in the cultivation of particular threatened plant species, with records carefully kept at the central garden, especially since in their skill and dedication such amateurs may often surpass their professional colleagues (Thomas, 1973). There is much to commend the suggestion of Mr Shaw, earlier in this symposium, that botanical gardens could perform a valuable conservation-orientated role merely by keeping track of the plants, especially the trees and shrubs, cultivated in their areas; the gardens themselves need not by any means attempt to grow all within their own walls. When it is impossible to preserve natural habitats in such a way as to ensure the survival of a given species, as when the plants are in imminent danger of destruction as by construction, strip mining, flooding, clearing, or heavy grazing, then the plants should clearly be removed to a botanical garden. The possibility of creating temporary reserves, where the plants can be maintained for a period of time until their cultural requirements have been investigated, is one option.

The five regions of the world with a mediterranean climate, although of limited extent, are inhabited by at least 25,000 plant species, of which more than half are endemic to a particular area. There are at least as many narrow endemics in regions with a mediterranean climate as on all the islands of the world put together, with the possible exception of New Guinea. Many of these are very narrowly restricted, and the creation of reserves may not be possible everywhere. Thus Moraea loubseri Goldblatt, a most attractive plant with blue-violet flowers about 5 cm. across, first found in 1973, appears to be endemic to one granite hilltop, Olifantskop near Saldanha Bay in South Africa. Several thousand plants are now in cultivation, but it will soon be extinct in the wild as its only habitat is being quarried for building stone and the hill will soon be gone. On the other hand, Calochortus tiburonensis A.J. Hill, a species with yellow flowers more than 3 cm. across, was discovered for the first time in 1972 on a hill overlooking San Francisco Bay and the single hilltop site where it forms a dense colony has been declared a reserve. In complex areas such as these, a system of relatively small reserves might provide one answer to the conservation of threatened plant species, although as already noted, this may not always be an option.

In the fynbos of the Cape Region, however, a formation that occupies some 1.8 million hectares of mostly mountain land, it is

largely the aggressive spread of alien trees and shrubs such as
Acacia, Hakea and Pinus, together with unseasonable burning, that
constitutes the greatest threat. As many as 1,500 species of
flowering plants of the Cape Region may be threatened with extinctio
(A.J. Hall, pers. comm.), and this is clearly one of the areas of
the world requiring the most urgent attention. It is suggested
that in this area especially careful management of the reserves
will be needed to ensure survival of the native species in view
of the explosive spread of the aliens. In the Mediterranean Basin
itself, on the other hand, there has been such a long history of
interaction with man and his animals that few relatively threatened
species may still exist, even though many of the narrow endemics
of the area may have reached a sort of equilibrium with environmental forces.

In view of the impressive concentrations of narrow endemics
they possess, areas with a mediterranean climate should be especiall
emphasized in conservation programmes, since here the rewards in
terms of numbers of species surviving are potentially great. In
addition, there are many botanical gardens located in these areas,
and an exchange of plants between them to be grown under controlled
conditions should be especially useful for conservation purposes.
Special care should be taken that these do not become weeds and
endanger the native flora of the areas to which they have been
brought.

On islands, introduced animals have often caused rapid and in
some cases irreversible changes in species composition, and many
of the species of islands are found nowhere else. In the Hawaiian
islands, for example, about half of the approximately 2,200 species
of native vascular plants are classified as threatened or of
uncertain status and 255 (11.6%) are already extinct (Anonymous,
1975a; Fosberg and Herbst, 1975). These figures compare with less
than ten per cent and one hundred species (about 0.5%) for the
continental United States. The cultivation of endemics from
seasonally dry areas at low elevations in Hawaii in botanical
gardens may be the only feasible way of preserving them at all as
these areas are rapidly being destroyed by urban and agricultural
pressures, whereas the endemics at higher elevations often grow in
regions of higher rainfall and are not therefore so susceptible to
human population pressure (W.L. Theobald, pers. comm.). In open
places at high elevations in Hawaii, grazing animals have very
serious effects, and it seems absolutely incomprehensible that the
authorities have thus far refused to exterminate the mammal
populations, apparently considering the sport that hunting them
provides to a few more important than the destruction and irre-
trievable loss of thousands of species of unique plants and
animals which evolved in the Hawaiian islands over millions of
years. The shifting relationships in natural communities on the
Galápagos Islands has recently been discussed by Hamann (1975).

In such areas as islands and other regions with very high
degrees of endemism, biologists would do well to consider whether
the local laws provide adequate protection to their unique biota.
The Canary Islands, for example, just such an area, are visited
annually by hundreds of thousands of tourists, some of whom make
extensive collections of plants or animals, including many which
are endangered. The flora of the islands is already well
represented in the herbaria at Kew, Paris and Florence, and one
might well call into question the scientific value of additional,
mostly random gatherings of these endemics. Would it not be
preferable to allow collecting and the export of plants only under
permit, still allowing scientific studies of all kinds when their
need had been established? It certainly would seem far more
important to attempt to provide future students with living
populations of these plants than with additional miscellaneous
herbarium specimens. Numerous other parts of the world probably
ought to consider additional legal protection for their wild plants
and animals, although the laws that might be enacted ought never
to be allowed to interfere with legitimate scientific studies.

Botanical gardens certainly might play an important role in
reintroducing rare and endangered species in natural habitats,
although this needs to be done with great caution, and only after
thorough documentation both of what is being introduced and what
the area in which the introduction is taking place was like before.
Sometimes even endangered species may assume unexpected roles in a
new habitat. Nor is it always possible to predict how a particular
species will react in the future, or under a new set of conditions,
as British experience with Epilobium angustifolium L., which has
expanded its range almost explosively since 1800, will amply
demonstrate.

In the Hawaii Volcanoes National Park in Hawaii, plants of the
very rare genus Hibiscadelphus are being propagated and then reintroduced into nature, and similar action might well be taken with
other genera such as Delissea and Tetraplasandra. We have already
mentioned the example of Teucrium scordium, which will be reintroduced into a locality from which it has disappeared in England.
Limonium tuberculatum (Boiss.) O. Kuntze, virtually exterminated
in the wild by the expansion of holiday centres along the southern
shore of Gran Canaria in the Canary Islands is being propagated in
cultivation. It is hoped that it soon will be re-established in a
small reserve within its native range (D. Bramwell, pers. comm.).
In the Canary Islands, also, only eight plants of Sideritis discolor
Webb are known in the wild, but the species is now well established
in cultivation, and efforts will soon be made to build up its
populations where it is native (D. Bramwell, pers. comm.). Similarly
D.R. Given (pers. comm.) has suggested that the very restricted
natural populations of New Zealand endemics Tecomanthe speciosa
W.R.B. Oliver (one plant) and Clianthus puniceus (G. Don) Sol. ex

Lindl. (fewer than twenty plants) be enriched by the addition of
additional established individuals from cultivation, where they
are frequent, especially since neither species is known to
reproduce in nature under present conditions. There may, however,
be difficulties. Seedlings of <u>Styrax platanifolia</u> Engelm. have
been planted out in or near native populations on the Edwards
Plateau of Texas, where it is endemic, but the seedlings rarely
survive, being eaten by deer and goats – the same reason that the
plant reproduces poorly in nature in the first place (M.C. Johnston,
pers. comm.). Additional protection, such as that accorded
<u>Ranunculus crithmifolius</u> subsp. <u>paucifolius</u> may be necessary for
this project to succeed.

In at least one case, a species seems to have become extinct
in the wild and then been re-established at a locality where it was
native. <u>Clarkia franciscana</u> Lewis & Raven, native to a single
serpentine covered hillside in the Presidio of San Francisco, was
actually thought to be extinct in the wild and was reintroduced
from cultivation in a botanical garden (Roof, 1972). It has
persisted subsequently and the resulting population has been
unchanged genetically for at least some alleles (Gottlieb, pers.
comm.).

In some cases the recreation of the natural habitat of a
particular plant species may be the only avenue possible for
conservation. Prairie communities are being artificially built up
in various places in the central United States, and it is reasonable
to expect that rare and endangered species would be introduced into
and maintained within these artificial communities. In addition,
certain plants can probably best be preserved by establishment in
suitable habitats where they were not previously found, since
reproduction under at least semi-natural conditions, with further
human intervention either unnecessary or limited, is clearly the
best alternative to the preservation of natural populations.

Thus D.R. Given has suggested that the New Zealand species
<u>Gunnera hamiltonii</u> T. Kirk, which is threatened at both of its
known localities, might be preserved by establishing it on one of
the reserved offshore islands; it is cultivated by rock gardens
and is an attractive plant. It should also be possible to re-
establish <u>Moraea loubseri</u>, mentioned above, on one or more of the
numerous granite hills in the Saldanha Bay area not threatened by
building activities (A.V. Hall, pers. comm.). Dr Hall has likewise
suggested that <u>Oxalis natans</u> L.f., the last remaining wild population
of which was recently destroyed in the wild, should be re-established
in some suitable natural habitat.

The establishment of a species extinct in the wild at a locality
similar to those it once inhabited and adjacent to them has been
accomplished at least once. In the San Joaquin Valley of California,

Cordylanthus palmatus (Ferris) J.F. Macbr. was exterminated in the
wild recently, but seems to have been re-established successfully
at a locality within the Mendota Wildlife Refuge, western Fresno
County, about 2 km. from a locality where it had just become
extinct (Anonymous, 1975a; L.R. Heckard, pers. comm.).

Such examples might well become more frequent as our options
inevitably become more limited, and our knowledge greater. They
might well be adopted in those cases in which a species is
threatened by some factor, such as grazing or disease, throughout
its entire natural range. The lessons taught us by the models of
island biology should, however, be kept in mind: the carrying
capacity of an area of given size under a particular set of
conditions is a constant, and the number of species in a reserve
will not in most cases be capable of elevation much above its
initial level over a period of time, although it might appear
initially that this has been the case. Either the additional
species or some that were initially present in the area will tend
to go extinct in time.

It is theoretically possible that genetically depleted and
much reduced natural populations might be reinvigorated by the
introduction of plants from other sources. A suggestion has been
made that this might be desirable for the rare South African
mountain protead Orothamnus zeyheri Pappe ex Hook., but in counter-
ing the suggestion, it was pointed out that the form of the species
with less attractive, dull bracts – and which is therefore less
desirable from a horticultural viewpoint – might prove dominant.
Doubtless, changes in the genetic constitution of wild populations
should be monitored with care, but again, survival is always
preferable to extinction whenever a clear case can be made.

To summarise this section, it may be said that the protection
of a species in natural or semi-natural surroundings where it can
maintain itself apparently offers the best hope for long term
survival at relatively low cost. Botanical gardens will increasingly become involved in activities of this sort as the range of
their conservation-related activities grows.

Turning now to the traditional role of botanical gardens,
specialized collections of particular groups certainly can be
developed, although history has shown that these rarely survive
the retirement or transfer of the individual who had the greatest
interest in them (Henderson, 1973b; Poppendieck, this symposium).
Heavy concentrations of plants of particular groups in one place
tend to invite their destruction by pests or diseases – one thinks
of Dutch elm disease, or lethal yellowing in palms. Furthermore,
there are very real difficulties caused by the possibility that
related species may hybridise in cultivation and thus fail to
maintain themselves. This may even be done deliberately, as in the

case of orchids, in which hybrids became so popular following the
development of successful techniques for growing these plants from
seed in the 1920's that many wild species disappeared from cultivation. As we have already mentioned, the enlistment of talented
amateur gardeners to form a sort of network around the "mother"
botanical garden could help enormously in the perpetuation of
large numbers of species in cultivation. Depending on how these
species were distributed, they could likewise be used as a way to
obviate the possibility of hybridisation between related species
and to maintain a greater range of genetic variability than would
be possible in one limited area of cultivation.

    For collections of threatened plant species in botanical
gardens, the normal problems involved in the conduct of such
institutions become all the more acute. Over a period of time,
labels do become mixed and records lost. If there is at least an
implied national or international responsibility for the maintenance
of certain collections, as Dr Poppendieck has pointed out in this
symposium, the plants might fare better. In the same vein, it is
suggested that long-lived plants such as trees and shrubs and
others which propagate themselves vegetatively will have much
better chances of survival in a botanical garden than annuals or
other short-lived species. As we have already mentioned and as
Dr Poppendieck has brought out in his remarks on the collection of
Mesembryanthemaceae at Hamburg, the long-term cultivation of
collections of plants for conservation purposes under glass is
probably a practical impossibility. Indeed, the most realistic
course for botanical gardens to take, if they wish to preserve the
maximum number of species, might be the creation of seed banks of
large scope (Hondelmann, this symposium), a course that is being
followed very successfully by the Royal Botanic Gardens, Kew, and
a few other institutions at present. This should be accompanied
by research on the most suitable conditions for seed storage. It
would certainly be appropriate for botanical gardens to produce
quantities of seed of rare and endangered species for storage in
a hoped for international network of seed banks once they have
established these plants in cultivation. They should also attempt
to disperse propagules of endangered species to as many centres as
possible to provide them with the maximum opportunity for survival,
but such a programme can probably be effective only with an international computerized system for retrieving information about the
plants cultivated in botanical gardens. An effort also should be
made to promote the maximum amount of genetic diversity in such
dispersal schemes. Tissue culture under appropriate conditions
might be even more economical of space than seed banks, but it
would as a method depend first upon the perfection of appropriate
methodologies, particularly those involved in regenerating plants
from tissue culture. Being highly energy dependent, both seed and
tissue banks require sustained funding at reasonable levels, and
are inferior to the preservation of natural or semi-natural

populations of living plants in this respect.

In closing this portion of the discussion, I would like to re-emphasize a theme which is mentioned several other times in this symposium. The cultivation of rare and endangered species in botanical gardens or in reserves or satellite gardens certainly offers advantages in advancing public understanding of the problem, and in generating support for the gardens themselves. On a world basis, however, botanical gardens are often supported for their display and amenity functions, and these may be more or less incompatible with the preservation of plant species. The kinds of reserves and specialized gardens in which rare plants are best preserved must have strictly limited public access, and the great majority of rare and endangered species are not particularly suitable for display in any case. Although, as Mr Shaw has suggested in this symposium, we might gain space for such unattractive rarities by limiting the degree of replication between existing botanical gardens, any such rearrangement must take into account the bases of support of the individual gardens, as well as their resulting display and educational value. In the broadest sense, whether botanical gardens will be able to respond effectively or not to the problem of preserving rare and endangered species of plants depends upon public acceptance of conservation as an important focal point of attention. Botanical gardens have a critical role to play in the public education that will continue to be basic in creating such acceptance. It may be that if they succeed in this effort, they will have created part of the rationale for their own continued existence, as Professor Heslop-Harrison has suggested.

THE COMMERCIAL USE OF WILD PLANTS

As recognized by the "Convention on International Trade in Endangered Species of Wild Flora and Fauna", signed at Washington, D.C., in March 1973, and ratified by fifteen countries as of 1 September 1975, a number of wild plants are being exploited commercially and merit international protection. The plants and animals listed in this convention may still be transmitted between scientific institutions in different countries by affixing an official label. In all other instances in which they are transported internationally, they will require a permit from the country of origin stating that their procurement and export has not further endangered the respective taxa in the wild. In the most strictly protected cases, listed in Appendix I in the Convention, a permit will also be required from the country of import, further certifying that this condition has been met. Among the plants which are commercially exploited at present are virtually all succulents and cycads as well as many orchids and bromeliads. In addition, such special cases as Dionaea, Darlingtonia, Panax, Dioscorea,

Harpagophyllum and Lodoicea may be mentioned. Many plants which
are collected in the wild for the horticultural trade appear to
merit special attention, especially since the rarest species are
often those particularly sought. For instance, this has been
brought out very clearly for North American cacti by writers such
as Weniger (1970) and Lyons (1972); at present, seventy-two of the
268 species of cacti found in the United States are considered
threatened.

        Another particularly dismal example is provided by the
exploitative collection of cycads and succulents in southern
Africa. These are collected not only for local use but especially
for amateur horticulturists in the United States, Europe and Japan.
In the Republic of South Africa, as in most other countries, plants
can be removed from private land for any purpose with the consent
of the landowner, and many of the rarest species will continue to
disappear from the wild unless such plants are given legal protection, as is the case in South West Africa, for example.

        Dr P.J. Wycherley, Director of the King's Park and Botanical
Garden in West Perth has pointed out (pers. comm.) that the
harvesting of cut flowers and other parts of native plants is a
severe threat to many species in Western Australia, an area very
rich in endemic taxa. Some of the commercially harvested species,
such as Macropidia fuliginosa (Hook.) Druce, Conospermum,
Lachnostachys and Verticordia appear to have a very low reproductive
capacity and to set seed only in certain years. Dr Wycherley has
estimated that although patches of Verticordia nitens Schau., for
example, may colour the ground golden for several kilometers,
perhaps only one in a million flower-heads may set viable seed.
In other species, such as Banksia laricina C.A. Gardner and
Dryandra polycephala Benth., undisturbed plants set seed abundantly
and regularly, but commercial exploitation of the flowers is so
great that seed set is drastically reduced in many areas year after
year. Combined often with burning at inappropriate times of year,
which is also a very serious problem in South Africa, such heavy
exploitation may cause even species which appear abundant to be in
fact endangered, as they may literally never be able to reproduce
themselves.

        Dr Wycherley has suggested that the collection from the wild
or from public lands for sale or export of any species of whose
survival status there is any doubt should be banned immediately,
and that in time the commercial exploitation of wild flowers growing
in nature should be prohibited. He suggests that there should be a
progressive changeover to the production of native wild flowers for
the trade by cultivation or to the management of private bushland
in a manner which will promote the survival of the species as well
as a crop of saleable material. Such usage should be encouraged
by suitable tax incentives, and botanical gardens could obviously

play a leading role in this whole area of research and development.

Despite a serious effort to investigate this point, very little
evidence was found of the deliberate reduction anywhere in the world
of populations of native plants to enhance the commercial value of
the remaining ones, a notorious practice of certain collectors in
the nineteenth century. Any examples of this sort that are discovered should be publicised; they certainly are repugnant to
most people.

Botanists can be serious offenders in the matter of destruction
of rare plants, as has become all too evident in heavily populated
countries such as Britain (Perring, 1975). In the tropics, it is
rarely possible to know whether collecting particular individuals
may put a given species at risk or not. On islands, however, where
there may often be only a few individuals of a particular species
or even genus in existence, botanical collecting should be carried
out with particular restraint, even if permitted by law, as it may
constitute an especially severe threat to the continued existence
of the plants concerned. The codes of conduct recently adopted by
the Botanical Society of the British Isles and by the International
Organisation for Succulent Plant Research for their members
represent excellent steps in the direction of self-policing by
concerned groups of botanists.

On the positive side, horticulture can play an enormously
important role in the preservation of rare and endangered plant
species, as it already has, for example, in the cases of
Ginkgo biloba L., Franklinia alatamaha Marsh., and
Encephalartos woodii Sander. One of the simplest, most effective,
and least costly ways in which botanical gardens can ensure the
perpetuation of particular species is to introduce them successfully
into the horticultural trade, or to encourage their wide use in
horticulture as discussed by Henderson (1973a) for Salix lanata L.
The efforts of the Provincial Department of Nature Conservation
in Grahamstown, South Africa, in propagating and distributing large
numbers of plants of various species of Encephalartos is an
excellent example. Making such plants available in quantity not
only helps to discourage gathering them in the wild, it also
contributes directly to the preservation of the species concerned
by spreading them widely in cultivation. The efforts of the
North Carolina Botanical Garden to develop methods for the efficient
propagation of such exploited plants as Dionaea and Sarracenia, as
well as other threatened species, are noteworthy in this regard.

Botanical gardens also could play a very useful function in
encouraging the use of rare and endangered species in plantings
such as those around public buildings and universities, where they
can serve an educational function and be maintained without continuing cost to the botanical garden (M.C. Johnston, pers. comm.;

Wallace, 1975). In the tropics, too, horticulturally useful plants
may be preserved in plantings; for example, mahogany,
<u>Swietenia humilis</u> Zucc., has been wiped out over most of the lowland
of Panama, but is one of the commonest street trees in the Canal Zon
It is of course a familiar practice of agriculturists everywhere to
preserve particular native plants which they consider important,
and this practice should be encouraged strongly by botanical
gardens.

   As suggested to me by J. Popenoe (pers. comm.), the tropics
undoubtedly contain the greatest reserve of unexploited horticultura
materials, many of which would make outstanding house-plants.
Especially in view of our limited knowledge about which tropical
plants are rare and endangered, it is clearly appropriate for
botanical gardens, both in tropical and in temperate regions, to
redouble their efforts to produce new tropical cultivated plants.
Rare orchids, for example, could be propagated by seed or meristem
culture and distributed widely, although computerized records like
those which Dr Hoffmann has begun would be necessary to ensure the
continued successful operation of the system as a whole. The
further development of horticultural industries based on native
plants in tropical countries might at once lead to the preservation
of many more species than is now likely, and at the same time
provide an economic incentive for doing so.

SUMMARY

   In the light of the preceding discussion, the major conclusions
of this paper may be summarised as follows:

1. Public education leading to an acceptance of the notion of
   preservation of threatened plant species is the foundation
   upon which all of our efforts must be built. Only with the
   long term financial commitment that would depend on such
   acceptance can a serious commitment be made to such preservation.

2. We must learn as well as possible which species are
   endangered and rare, and then construct international
   computerized systems which will tell us which are in
   botanical gardens and reserves, or on public lands, at
   present. For the continental United States it is estimated
   about two-thirds of the threatened species are growing on
   public lands (E.S. Ayensu, pers. comm.). The Plant Records
   Centre (PRC) of the American Horticultural Society already
   incorporates about 135,000 computerized records of plants
   grown at many different gardens, and might provide an
   excellent base on which international co-operation in this
   area might be built.

3. Conservation *in situ* is to be preferred to any other method, with conservation by the establishment of spontaneously reproducing colonies in natural environments next best.

4. Research should be carried out actively on the proper methods of propagating rare and endangered species.

5. When choices must be made between species, or priorities must be assigned, first choice should be given to plants of actual or clearly demonstrated potential use in fields such as agriculture, forestry, horticulture and pharmacology, and for scientific research.

6. A short list of rare and endangered plants of high scientific importance should be set up, and botanical gardens should collectively accept the responsibility for bringing them into cultivation under circumstances in which they could reasonably be expected to survive in cultivation indefinitely. The list might include such genera as *Cathaya*, *Degeneria*, *Gomortega*, *Indiospermum*, *Lactoris*, *Malagasia* and *Sarcopygme*, to name but a few. The acceptance of such a list as a common goal for the botanical gardens of the world would help to lay the foundation for further collective efforts, be most useful in arousing public interest, and lead to the preservation of the plants involved.

7. Plants should in general be maintained where they can be grown out-of-doors, without high costs in energy and the concomitant risk of losing the plants through power failure. Botanical gardens should form a network within which materials would routinely be transferred to climates in which they would grow well, as in the example of sending Somalian plants to Adelaide mentioned by Mr Simmons in this symposium. Possibly replicates of Professor Ihlenfeldt's outstanding collection of *Mesembryanthemaceae* could be transferred to South Africa, as well as to southern France, southern California, and South Australia when it is no longer required in Hamburg.

8. Except for instances of special economic or scientific importance, we should abandon the effort to pick out the most threatened species among the roughly 155,000 found in the tropics, and concentrate instead on general floristic surveys, conservation, horticulture and helping tropical scientists and institutions to do the job for themselves. At least 50,000 tropical species will reach threatened status or become extinct by the end of this century with the greatest number in Latin America; there is no way in which a large number of these can be recognised as

endangered species and brought into cultivation before they reach this status.

9. In temperate regions, with a combined flora of about 85,000 species, about 4,500 appear to be threatened at present. Many of these are in regions of mediterranean climate, with at least 12,000 local endemics, and such regions should be given special emphasis. About 1,500 of the threatened species are in the Cape Region of South Africa alone. Islands also should be treated individually. As many as possible of the 4,500(-6,000) threatened species in temperate areas should be brought into cultivation, distributed actively to botanical gardens in climatic zones where they will thrive - the US and USSR have recently signed a binational agreement for collaboratio in this effort - and monitored indefinitely. They should also be accumulated as rapidly as possible in seed banks. First priority should of course be given to those species not reproducing on public lands, and to those of out-standing scientific or economic importance. Only seeds and cuttings of such threatened species should be gathered in the wild unless their destruction is virtually certain, in which case transplants might be attempted. We certainly have the capability to preserve most of the flora of the temperate zone for posterity through a combination of reserves and cultivation in botanical gardens, and I hope that most of us will share my belief that we should organise here at Kew and begin to make preparations to do just that.

REFERENCES

[Anonymous]. (1970). Guidelines for biological field work. Science 169: 8.

[Anonymous]. (1974). Trends, priorities, and needs in systematic and evolutionary biology. Systematic Zoology 23: 416-429.

[Anonymous]. (1975a). Report on the endangered and threatened plant species of the United States. U.S. Government Printing Office, Washington, D.C., (Serial No. 94-A).

[Anonymous]. (1975b). The use of ecological guidelines for development development in the American humid tropics. I.U.C.N. Publications (new ser.) 31: 1-249.

BENSON, L. (1975). Conservation of cacti. National Parks and Conservation Magazine (in press).

BERNARDI, L. (1974). Problèmes de conservation de la nature les
    îles de l'Océan Indien. 1. Méditation à propos de Madagascar.
    Saussurea 5: 37-47, 4 pl.

BUDOWSKI, G. (1974). Scientific imperialism. Unasylva 27: 24-30.

BUDOWSKI, G. (1974). Forest plantations and nature conservation.
    I.U.C.N. Bulletin 5: 25-26.

CASEY, T.L.C. and JACOBI, J.D. (1974). A new genus and species of
    bird from the Island of Maui, Hawaii (Passeriformes:
    Drepanididae). Occasional Papers of Bernice P. Bishop Museum
    24: 215-226.

DIAMOND, J.M. (1972). Biogeographic kinetics: estimation of
    relaxation times for avifaunas of Southwest Pacific Islands.
    Proceedings of the National Academy of Sciences of U.S.A.
    69: 3199-3203.

DIAMOND, J.M. (1975). The island dilemma: lessons of modern bio-
    geographic studies for the design of natural preserves.
    Biological Conservation (in press).

ELTON, C.S. (1975). Conservation and the low population density
    of invertebrates inside Neotropical rain-forest. Biological
    Conservation 9: 3-15.

FARNWORTH, E.G. and GOLLEY, F.B. (1974). Fragile ecosystems.
    Springer-Verlag, Berlin.

FISHER, F.J.F. (1965). The alpine Ranunculi of New Zealand.
    Bulletin of the New Zealand Department of Scientific and
    Industrial Research 165: 1-192.

FOSBERG, F.R. and HERBST, D. (1975). Rare and endangered species
    of Hawaiian vascular plants. Allertonia 1: 1-72.

FOWLIE, J.A. (1975). Malaya revisited. Part VI.
    Paphiopodilum godefroyae in the Birdsnest Islands of Chumphon,
    east of the Isthmus of Kra. Orchid Digest 39: 32-37.

FRANKEL, O.H. (1974). Genetic conservation: our evolutionary
    responsibility. Genetics 78: 53-65.

GÓMEZ-POMPA, A. and A.BUTANDA, C. (eds.)..(1975). Index of current
    tropical ecology research, Vol. 1. Instituto de Investigaciones
    sobre Recursos Bióticos, México, D.F.

GÓMEZ-POMPA, A., VÁZQUEZ-YANES, C. and GUEVARA, S. (1972). The
    tropical rain-forest: a non-renewable resource. Science
    177: 762-765.

HAFFER, J. (1969). Speciation in Amazonian forest birds. Science
     165: 131-137.

HAMANN, O. (1975). Vegetational changes in the Galapagos Islands
     during the period 1966-73. Biological Conservation 7: 37-59.

HENDERSON, D.M. (1973a). The role of botanic gardens in horti-
     culture and field botany. In GREEN, P.S. (ed.), Plants: wild
     and cultivated: 66-71. Botanical Society of the British Isles.

HENDERSON, D.M. (1973b). [Discussion]. In GREEN, P.S. (ed.),
     Plants: wild and cultivated: 58. Botanical Society of the
     British Isles.

HESLOP-HARRISON, J. (1973). The plant kingdom: an exhaustible
     resource? Transactions of the Botanical Society of Edinburgh
     42: 1-15.

JANZEN, D.H. (1973). Tropical agroecosystems. Science 182: 1212-
     1219.

LYONS, G. (1972). Conservation: a waste of time? Cactus &
     Succulent Journal (U.S.A.) 44: 173-177.

MAY, R.M. (1975). Island biogeography and the design of wildlife
     preserves. Nature 254: 177-178.

NICHOLLS, Y.I. (compiler). (1973). Source book: emergence of
     proposals for recompensing developing countries for maintain-
     ing environmental quality. I.U.C.N. Environmental Policy and
     Law Paper 5: 1-143.

PERRING, F.H. (1975). Problems of conserving the flora of Britain.
     Botanical Society of the British Isles News 9: 18-24.

PRANCE, G.T. (1973). Phytogeographic support for the theory of
     Pleistocene forest refuges in the Amazon Basin, based on
     evidence from distribution patterns in Caryocaraceae,
     Chrysobalanaceae, Dichapetalaceae and Lecythidaceae. Acta
     Amazonica 3(3): 5-28.

RATCLIFFE, D.A. (1973). Safeguarding wild plants. In GREEN, P.S.
     (ed.), Plants: wild and cultivated: 18-24. Botanical Society
     of the British Isles.

ROOF, J.B. (1972). Notice of Clarkia franciscana sowings. Four
     Seasons 4: 17.

SPECHT, R.L., ROE, E.M. and BOUGHTON, V.H. (eds.). (1974).
    Conservation of major plant communities in Australia and
    Papua New Guinea. Australian Journal of Botany (Suppl. Ser.)
    7: 1-666.

TERBORGH, J. (1974). Preservation of natural diversity: the
    problem of extinction prone species. Bio-Science 24: 715-722.

TERBORGH, J. (1975). Faunal equilibria and the design of wildlife
    preserves. In GOLLEY, F.B. and MEDINA, E. (eds.), Tropical
    Ecological Systems, (Ecological Studies, Vol. 11): 369-380.
    Springer-Verlag, New York, Heidelberg, Berlin.

THOMAS, G. (1973). The role of the private garden. In GREEN, P.S.
    (ed.), Plants: wild and cultivated: 59-65½ Botanical Society
    of the British Isles.

TINKER, J. (1974). Why aren't foresters more conservationist?
    New Scientist 61: 819-821.

VANZOLINI, P.E. (1970). Zoologia sistemática, geografia e a origem
    das espécies. Instituto de Geografia, Universidada São Paulo,
    Série Teses e Monografias 3: 1-56.

WALLACE, G.D. (1975). Suggestions for preventing extinction of rare
    and endangered plants. Arboretum and Botanical Garden Bulletin
    9: 67-68.

WALTERS, S.M. (1971). Index to the rare endemic vascular plants of
    Europe. Boissiera 19: 87-89.

WENIGER, D. (1970). Cacti of the Southwest. University of Texas
    Press, Austin, Texas.

WETZEL, R.M., DUBOS, R.E., MARTIN, R.L. and MYERS, P. (1975).
    Catagonus, an "extinct" peccary, alive in Paraguay. Science
    189: 379-381.

DISCUSSION SECTION VI TECHNIQUES OF COLLECTING

Professor Hedberg (Uppsala) said that most of the effort appears to be directed at temperate areas, but the tropics need to be considered. The first priority is to know what must be conserved and in order to do this the description and collection of such material must be undertaken as quickly as possible. Botanic gardens can help to make the public understand the necessity of establishing reserves for endangered plants. The Conference should recommend the setting up of botanic gardens in tropical countries such as Ethiopia where they do not exist at present.

Professor Zohary (Jerusalem) pointed out that when gene pools of wild or cultivated plants are sampled the periphery of their distribution should not be neglected because in such peripheral areas varieties with important traits like disease resistance may be found.

Sir Otto Frankel (Canberra) doubted that this was true for annual crops. He agreed with a point made by Professor Hawkes that some elaborate plant collecting forms are designed more for the computer than for real botanic usefulness. An important additional column on such a form would be one for data on associated plant groups.

Professor Zohary (Jerusalem) stressed that many ancient cultivars of crop plants and their wild relatives have a distribution pattern with a massive central area and a number of scattered peripheral populations. The latter are much more difficult to locate but contain plants with useful genes. Such genotypes have become adapted to conditions not found in the central area of distribution.

Professor Hawkes (Birmingham) agreed that it is useful to sample the edges of a distribution range where gene pools are under environmental stress. This may yield alleles of value to plant breeders but it must not be forgotten that equally important alleles are also present in the central area.

Professor Böcher (Copenhagen) said that biotypes of rare species are difficult and costly to conserve and the public should be made aware of the necessity of protecting small areas of the native habitat of such plants. The ease with which this work can be carried out depends upon the species involved. The public are easily convinced if an orchid is threatened, but not if it is an unattractive plant.

Mr Hepper (Kew) reported that in Northern Tanzania cycads appear to be endangered by commercial interests. He added that many tropical botanic gardens were established a century or more ago in association with Kew or other European gardens. Recently their work and continued existence has been subject to financial and political pressures. He suggested that an international agency might be able to re-establish such connections to the mutual benefit of all.

Professor Gomez-Campo (Madrid) said that when plants are collected, data on their phytosociological association and also on the interaction of climatic and edaphic factors should be recorded.

Dr Stearn (British Museum, Natural History) underlined the importance of the amateur gardeners and plant collectors in conservation and suggested that botanic gardens should share seeds of rare species, not only with other gardens, but with private individuals who are known to have an interest and ability in growing specialist groups.

Dr Melville (Kew) said that since the tropical rain forest is one of the most highly endangered habitats there is a great need for botanic gardens in the tropics to undertake conservation projects. Game parks are very successful in the preservation of wild animals, and he suggested the establishment of floral parks of a suitable size for plants. Some of the area could be land-scaped and planted with some of the more spectacular and beautiful native plants while the major part could be set aside as a wilder-ness garden, not open to the ordinary visitor, where research into conservation could be undertaken. He hoped that these would be self-supporting and enhanced by introduction of animals which can be controlled carefully in relation to the plants.

Functions of
Conservation-Orientated
Collections in Research

GENETIC FACTORS TO BE CONSIDERED IN MAINTAINING LIVING PLANT COLLECTIONS

Karl Esser

Ruhr-University

Bochum, Federal German Republic

SCOPE OF MAINTAINING PLANT COLLECTIONS

At a time when public opinion has become focussed on environmental problems and the protection of wild life, the maintenance of living plant collections has also assumed considerable importance. Maintenance does not only involve economic and technical problems but also problems of basic research mainly relating to genetics. This brings up the question: What do we want to maintain and what can we maintain? There are two alternative possibilities to consider:

1. To keep the populations concerned as far as possible under the same conditions as they grow in their natural environment so as to allow them to follow the same lines of evolution as in nature; this may be called conservation.

2. To keep the populations at the stage of evolution at which they were taken from nature; this may be called preservation.

In reflecting on these possibilities, it becomes obvious that conservation in the above sense cannot be realized, for the following reasons:

1. Despite being able to imitate some environmental conditions (eg. light, temperature) one can never restore the whole spectrum of ecological conditions, especially since we are dealing in most cases with small isolated populations of single species and not with biotypes.

2. Due to sexual reproduction, the genotypes are submitted to a continuous alteration by recombination of the genetic material. As a consequence of the incompleteness of environmental conditions, selection is out of control and will follow different lines from those in nature.

We are left therefore with the second alternative. Thus the aim of maintaining collections can only be to preserve as much as possible the genotypes of the plants as they were taken from nature, knowing that whatever method we use to maintain plants, this will not be equivalent to their maintenance in a natural population.

WAYS OF MAINTAINING PLANT COLLECTIONS

There are two ways of maintaining plant collections:

1. Vegetative propagation.

2. Sexual reproduction.

In considering the advantages and disadvantages which are offered by both methods we shall come to realize the importance of both genetic and physiological parameters.

1. Vegetative propagation is the usual method used to keep strain collections of lower organisms. It can also be applied to trees, shrubs and perennial plants and even annual herbs may be conserved by cuttings. However, with large collections the latter is a very laborious task which in practice is almost impossible on a large scale. Vegetative propagation has the advantage that the original genotype is preserved as far as possible, but in doing this on a long term scale the expression of the natural variability, which is required for survival, is suppressed. This disadvantage compensates to a large extent the benefit obtained by "genetic stability".

In this context I should like to mention the negative experience gained from the maintenance of microbial collections. It has been found that strains kept alive entirely by vegetative transfers sooner or later become senescent and die due to an accumulation of negative mutations in the chromosomal or extrachromosomal genetic material.

This may also occur in herbs which are continuously propagated by cuttings, despite the fact that their genomes may compensate for defective mutations for a longer time than the haploid genomes of micro-organisms. Another very well-

known example that may be quoted is the loss of viability
in potatoes after long vegetative propagation and the concomitant increase in sensitivity to virus.

2. Sexual propagation by seeds is the usual means of maintaining annual plants. It has the practical advantage that
many seeds can be stored for longer periods (up to several
years) and can be transferred to other locations more
easily than growing material. However, since most plants
taken from nature are to a very high degree heterozygous,
one will obtain a high variability even in the first
generation.

This variability is caused by recombination which is the
result of karyogamy and meiosis. Karyogamy, where male and female
gametes or nuclei fuse, provides the basis for meiosis where
recombination takes place. It is not important that in diploid
higher plants these two events are separated by the whole vegetative development of the plant. It may sound trivial, but it is
essential to stress the fact that the basis for meiotic recombination has already been laid by the foregoing karyogamy. All the
conditions therefore which control karyogamy control the efficiency
of recombination. The main factors which determine whether
karyogamy takes place or not are genetic; they decide whether
fusion is between highly heterogenic gametes or between homogenic
gametes, and they thus control the effectiveness of recombination
in leading to descendants with altered genetic information or
unaltered genetic information.

These genetic factors act through the framework of the so-
called breeding systems. Breeding systems control the recombination
of genetic material. From this it follows that in order to maintain plants by seed we have to understand the factors which control
breeding systems and therewith sexual reproduction. If we wish to
limit variability we must learn how to manipulate these factors.

GENETICS AND SIGNIFICANCE OF BREEDING SYSTEMS

In Figure 1 we have a scheme showing the main breeding systems,
and indicating their action and interaction. From this figure we
may deduce the following:

1. Monoecism (hermaphroditism) constitutes the simplest
system. Female and male sex organs are formed on the same
plant. There is no restriction whatsoever on self- or
cross-fertilization. In contrast to animals, this system
is predominant in higher plants. It makes no difference
whether the sex organs are localized in one hermaphrodite
flower or whether they occur in male or female flowers

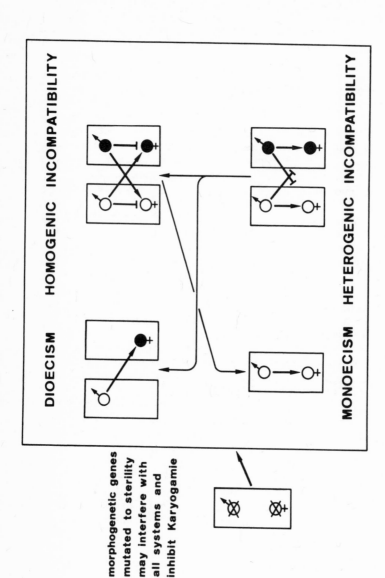

Figure 1. Action and interactions of breeding systems in higher plants. The rectangles in the region of different systems represent single individuals of one race. The rectangles concerning heterogenic incompatibility are an exception to this insofar as they symbolize single races or even species. The sexual symbols indicate female and male gametes. Differences in the genetic information of the gametes are marked by black and white. The bold arrows indicate the direction of karyogamy. The blocked arrows indicate the impossibility of karyogamy. Interactions between various systems are indicated with thin arrows (from Esser 1971, modified).

(eg. Cucurbita), self-fertilization dominates and thus favours inbreeding. Monoecious self-fertilizers are therefore predominantly homozygous and variability amongst the offspring is low. As regards the genetic control of this system, there are no special genes besides those determining the morphogenesis of sex organs.

2. Dioecism, where the female and male sex organs are formed on different plants, naturally requires obligate cross-fertilization. This system favours outbreeding and is rare in higher plants (only ten per cent of genera, see Heslop-Harrison, 1972). Dioecious plants are extremely heterozygous and the variability in their offspring is high. The genetic control of dioecism is thought to lie mainly with sex chromosomes, but the possibility that single genes are responsible for the sexual differentiation into male and female plants in some species is not excluded.

3. Permanent self-fertilization within monoecious (hermaphrodite) plants could cause an inbreeding depression and lead over the long term of evolution to an elimination of the species concerned. However, inbreeding is commonly prevented by a specific system which is widespread only in plants (about 3,000 species out of fifty-eight families). This is incompatibility, which may be defined in a very general way as the inability of normal functional gametes to fertilize each other. This barrier to seed production is caused genetically by special traits which are different from those responsible for sexual differentiation or dioecism. The best known system becomes effective when the gametes have like genetic determinants. We have therefore termed it homogenic incompatibility. Homogenic incompatibility is always connected with self-incompatibility and cross-compatibility. The latter occurs only between plants having different incompatibility factors.

Homogenic incompatibility as a phenomenon has been known for more than 200 years. As early as 1764, Kölreuter observed "Selbststerilität" in Verbascum and it became better known with Darwin's work. Since then numerous different mechanisms have been detected (for literature see review of Linskens and Kroh 1967, Arasu 1968, Townsend 1971) which differ in the physiological action and the sites of inhibition. They all have the main feature in common: the inability of genetically like determined gametes to fuse. This inhibition may come about at various stages of the fertilization procedure, eg. inhibition of germination of the pollen grains on the stigma; inhibition of pollen tube growth in the style; inability of the pollen tube to enter the ovule.

In order to demonstrate this system I would like to mention two main groups: the homomorphic plants, the flowers of which have pistils and stamens of the same size and the heteromorphic plants where the flowers of each individual differ in the length of pistils and stamens.

Where the genetic mechanism is concerned, we can also distinguish two types: the gametophytic mechanism, where each incompatibility gene acts independently in the haploid pollen tube and the diploid stylar tissue and the sporophytic mechanism where, in contrast, the gene action of the pollen tube and stylar tissue respectively is determined by the diploid sporophyte.

Complicated as nature is, it is not surprising that incompatibility mechanisms exist combining the above four parameters: homomorphy, heteromorphy, gametophytic and sporophytic determination. Thus it becomes evident that an understanding of the breeding system present in a species needs careful investigation for each individual case.

In order to underline this statement we may consider two examples of homogenic incompatibility in higher plants:

1. Homomorphy correlated with gametophytic determination is a rather common mechanism which occurs in many herbs and trees (eg. Solanaceae, Rosaceae). As may be seen in Petunia (Figure 2), the fertilization procedure is controlled by the so-called S-locus, the alleles of which show no interaction. Incompatibility occurs when pollen tubes grow in a style having at least one S-allele in common. Compatibility only exists when pollen tubes grow in a style having different S-alleles.

2. Heteromorphy correlated with sporophytic determination is less common (eg. Primula, Forsythia, Fagopyrum). As may be seen from Figure 3, in Forsythia two types of plants exist differing by the length of their styles and stamens respectively, called long-styled (pin) and short-styled (thrum). This morphological difference is controlled by a single locus having two alleles A and a. Long-styled plants are homozygous a/a and short-styled plants are heterozygous A/a. Concomitantly with the determination of flower morphology, the A and a genes are also responsible for the fertilization procedure. Due to the dominance of A the A/a short styles allow normal growth of all a pollen tubes originating from the long-styled a/a plant. However, the a pollen of the short-styled plant behaves phenotypically like A and is only compatible with the a/a long style and incompatible with its own A/a style.

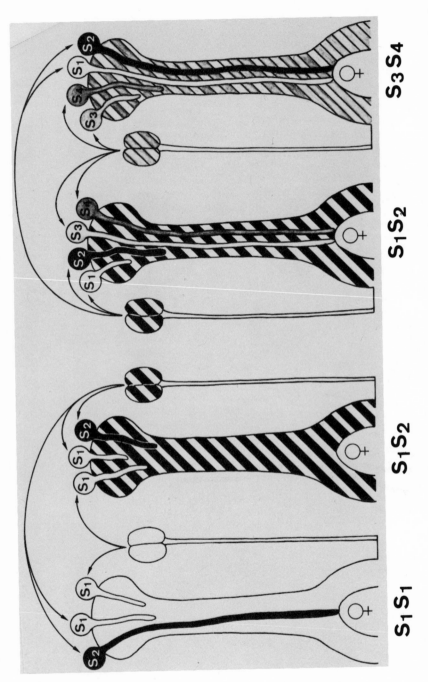

Figure 2. Genetic control of homogenic incompatibility in homomorphic plants by the gametophytic mechanism, eg. Petunia. In the fertilization procedure each allele acts independently. Incompatibility occurs when the 'S' allele of the pollen tube is present at least once in the stylar tissue.

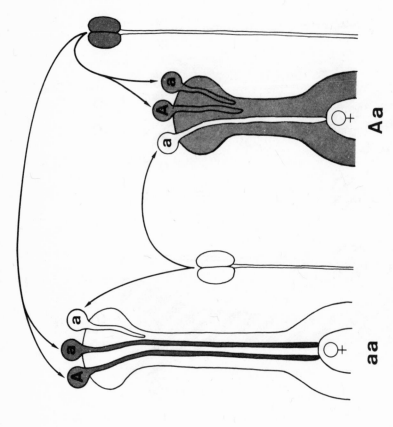

Figure 3. Genetic control of homogenic incompatibility in heteromorphic plants by the sporophytic mechanism, eg. Forsythia. In the fertilization procedure the pattern of pollen tube growth is determined by the sporophyte. All pollen grains and also the stylar tissue of the heterozygous (Aa) short-styled plant behave phenotypically like the dominant gene A. Therefore a pollen of the Aa plant is incompatible with its own style and compatible with the aa long-styled plant.

This phenomenon of predetermination by the sporophyte was
described by Darwin. However, its physiological implications have not been understood until quite recently, as
shown by the work of Professor Heslop-Harrison and his
collaborators.

3. In addition to these so-called dimorphic plants, there are
   other heteromorphic plants, like Lythrum and Oxalis, which
   have three types of flowers. These trimorphic types have
   the sex organs on three levels, ie. long styles with mid
   and short stamens; mid styles with long and short stamens;
   short styles with long and mid stamens. There are two loci
   controlling both the flower morphology and the sporophytic
   determination. In stating that the fertilization procedure
   follows the same rule as in dimorphic species, ie. compatibility occurs only between sex organs of equal length, we
   can avoid discussing further details of the rather complicated genetics of the trimorphic plants.

   In concluding the discussion of homogenic incompatibility
   we must mention its effect on variability. As can be seen
   from Figure 1, this breeding system overlaps monoecism.
   It leads phenotypically to the same effects as dioecism,
   because it requires for seed production genetically
   different plants and as a consequence promotes hetero-
   zygosity and thus variability. Homogenic incompatibility
   therefore occupies the same position in the evolution of
   plants as dioecism does in the animal kingdom.

4. In addition to these three breeding systems there is a
   fourth system which has only come to our attention quite
   recently. This is heterogenic incompatibility, a system
   which was first genetically analysed by us some twenty
   years ago in a fungus, the ascomycete Podospora anserina
   (for literature see Esser and Blaich 1973). Heterogenic
   incompatibility is, as the term says, determined by the
   inability of genetically different gametes to fuse. The
   phenomena of pollen tube inhibition are the same as in the
   homogenic system. Heterogenic incompatibility has never
   been found within the members of one race, it occurs only
   between different races. It overlaps monoecism as well as
   dioecism. It may also occur between races which in intra-
   race crosses are controlled by the homogenic system.

   In higher plants heterogenic incompatibility shows up in
   most cases as an unilateral incompatibility, ie. in
   reciprocal crosses between plants of different races (even
   different species). In one cross there is normal seed
   production, whereas in the other cross the pollen tubes are
   inhibited in the style. The mechanisms of genetic control

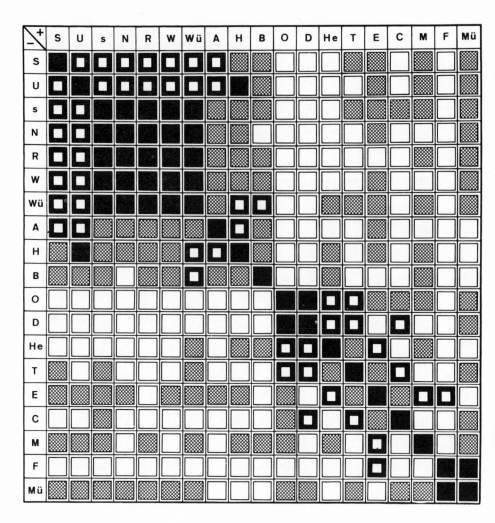

Figure 4. Scheme of the mating reactions between various races of
the fungus <u>Podospora anserina</u> isolated from different
localities in Germany and France. Meaning of the symbols:
+/- mating type; capital letters = designation of races;
black squares = compatible in both vegetative and sexual
phase; black squares with inserted open squares = sexual
compatibility and vegetative incompatibility; hatched
squares = vegetative incompatibility and reduced fruit
body formation (partial sexual incompatibility); open
squares = incompatibility in both vegetative and sexual
phase (adapted from Esser, 1971).

are manifold, eg. the S-genes of the homogenic system;
additional factors to the S-genes, and factors other than
S-genes. The physiological mechanism of this system is not
understood yet, but in all cases the block in fertilization
is caused by genetic differences and not by genetic identity
as in the homogenic system.

Heterogenic incompatibility has a similar isolating effect
to geographical barriers inhibiting the gene flow (for
literature see Ehrlich and Raven 1969). Thus it has a
restrictive effect on variability (Figure 4). This
decrease in variability depends, however, on which of the
three other breeding systems is overlapped by this system.

Heterogenic incompatibility occurring concomitantly with
monoecism has the most drastic effect, whereas in combination with dioecism or homogenic incompatibility it counter-
acts the enhancing effect on the variability promoted by
these two systems.

To summarise: there are four breeding systems controlling
in different ways karyogamy and thus recombination, hetero-
zygosity and variability. Their counteractions and overlaps
have caused the many difficulties encountered in understanding the genetic control of sexual reproduction in
higher plants.

                    OTHER GENETIC PARAMETERS

     In addition to the breeding systems there are more genetic
factors which have to be considered for the preservation of plant
collections.

1. Sterility genes. In higher plants the morphology and
   function of sex organs or gametes is controlled by numerous
   morphogenetic genes. Mutations occurring at these loci,
   which are not identical with the incompatibility factors,
   may lead to malfunction of these genes and thus cause a
   lack or functional inability of the gametes concerned.
   Morphogenetic genes mutated to "sterility" or, shortly,
   "sterility genes" may cause great confusion in understanding which, or how many breeding systems are effective in a
   plant (see Figure 1). This is especially the case when
   sterility genes cause no obvious morphological effects,
   only physiological defects. If self- or cross-fertilization
   of a plant is not possible, one has first to check whether
   this is due to sterility or incompatibility genes.

2. Chromosome mutations like translocations, deletions or inversions may lead to irregularities during meiosis and may cause the development of non-functional gametes which cannot be distinguished from those produced by the presence of sterility genes.

3. Genome mutations like aneuploidy and polyploidy may have similar effects during meiosis leading also to production of sterile gametes.

From these brief statements it follows that a controlled manipulation of sexual reproduction requires both genetic and cytological studies.

CONCLUSIONS

This presentation of the genetic parameters instrumental in sexual reproduction brings up the question of practical application. How can the results of basic research have practical application in maintaining collections of living plants?

One may try as much as possible to maintain plants by vegetative propagation, as mentioned above, and take the risk of loss of viability.

If sexual reproduction has to be used to maintain a collection one needs above all a rather large population in order to have a sufficiently large gene pool to preserve the population. There is no doubt that variability cannot be excluded. In this case it is of paramount importance to be able to manipulate this variability in order to profit from its positive effects and to avoid its negative effects. This once more requires a knowledge of the breeding systems and other genetic factors.

This involves also a knowledge of practical manipulation, eg. it may be necessary to produce seeds from a self-compatible plant. This can be done in making use of the phenomena of bud fertilization or of end-season fertility (seed production at the end of the flowering period). Both methods have come to our attention by experience and how they overcome the incompatibility barrier is not yet fully understood.

One other practical hint: Would it not be a good idea to mark in the seed catalogues or in any plant lists, in addition to the authors who have established this species, all the known genetic characteristics like incompatibility genes, etc. as is done in microbial strain collections.

In making controlled pollinations and comparing the offspring with its parents, one is not able to preserve the genetic information of each individual but one may be able to keep the genetic information of the population as a whole. In following this line, while on the one hand recessive lethal genes will be eliminated, on the other a certain "domestication" will occur, because one can never phenocopy all environmental conditions when plants are kept in "prison". This domestication may result in a loss of genes responsible for resistance.

In the introduction we have already stated that a conservation of plant collections in botanic gardens is not possible and it follows from this discussion of genetic problems that a true preservation is also difficult. This should not discourage, indeed, in facing this truth we have really made a step forward, because nothing which occurs is likely to be unexpected. In knowing the boundaries and in being able to channel plant conservation on the basis of the knowledge and application of genetic parameters, we are at least aware of our limits.

Knowledge of limits means that even in using these genetic parameters we cannot predict what may and what will happen with a population. It is analogous to a traffic lights system, you can channel the traffic in certain directions but you cannot avoid accidents altogether. We can channel the gene flow but never avoid "accidents" in plant populations.

In concluding I would like to bring to your attention a thought: It has always been the aim of science to question itself, and indeed there is no real progress in science without permanent self-criticism. This leads to the crucial question: Why do we keep collections of living plants whilst knowing that we can never conserve or preserve them as nature does? If we cannot answer this question, maintaining plant collections may become an end in itself without any scientific or economic value, like keeping pictures in a museum or animals in a zoo. But I think we can answer this question: Maintaining collections of living plants has the following justifications: (1) Preserving a gene reservoir for basic and applied research, eg. breeding purposes; (2) conserving rare plants which may otherwise die out; (3) conserving plants whose economic potential has not yet been investigated; (4) last but not least using these plants for educational purposes.

REFERENCES

ARASU, N.T. (1968). Self-incompatibility in angiosperms: a review.
    Genetica 39: 1-24.

EHRLICH, P.R. and RAVEN, P.H. (1969). Differentiation of populations
    Science 165: 1228-1231.

ESSER, K. (1971). Breeding systems in fungi and their significance
    for genetic recombination. Molecular and General Genetics
    110: 86-100.

ESSER, K. and BLAICH, R. (1973). Heterogenic incompatibility in
    plants and animals. Advances in Genetics 17: 107-152.

HESLOP-HARRISON, J. (1972). Sexuality of angiosperms. In
    STEWARD, F.C. (ed.) Plant physiology: a treatise. Vol. VIC:
    Physiology of development: from seeds to sexuality: 133-289.
    Academic Press New York and London.

LINSKENS, H.F. and KROH, M. (1967). Inkompatibilität der Phanero-
    gamen. In RUHLAND, W. (ed.) Handbuch der Pflanzenphysiologie
    Vol. XVIII: 506-530. Springer Berlin, Heidelberg and
    New York.

TOWNSEND, C.E. (1971). Advances in the study of incompatibility.
    In HESLOP-HARRISON, J. (ed.) Pollen: development and
    physiology: 281-309. Butterworth, London.

REPRODUCTIVE PHYSIOLOGY

(Transcript)

J. Heslop-Harrison

Royal Botanic Gardens

Kew, Surrey, England

In my opening remarks to the conference I noted that the ideal policy for the indefinite conservation of plant gene pools is undoubtedly the establishment of reserves of adequate size to maintain samples of the important ecosystems in conditions where the natural population dynamics of all the flora and the dependent fauna are impaired as little as possible. The ideal is not one that can now be easily implemented, since it is questionable whether undisturbed samples of appropriate size can always be found. Experience in many parts of the world is showing that conservation in the field demands much more than simply putting a fence around a fragment of a particular association and leaving it to its own devices. As the areas of the major forest, grassland and other ecosystems diminish, so we find that positive management of the residues becomes more and more essential. Such management will often demand special attention with regard to establishment and regeneration, not only of the dominants, but of the whole associated flora if the aim is to preserve something near a realistic sample of the original balanced association. This is especially true for forests. It is often asserted, probably quite correctly, that timber extraction does not necessarily have to destroy the original forest community structure completely, nor to eliminate the associated non-arboreal flora, provided that exploitation policies are adopted which permit natural reseeding and re-establishment. Yet the fact is that we lack much of the fundamental knowledge needed to formulate such policies. We are lamentably ignorant of the reproductive systems of all but a tiny fraction of the angiosperms, and for the tropics, where the problems are most urgent, the proportion is the least. Some of the gaps in knowledge can be filled by the use of living collections and it is my purpose in this paper to point out a few of the opportunities.

It must first be said that many aspects of plant reproduction can be satisfactorily studied only in the field. This is true notably of pollination and seed dispersal mechanisms, where the individual plant or plant species functions, so to speak, as one constituent of the greater "organism", namely the biome as a whole. Many instances are already known where the failure of regeneration of a species has been traced to a disruption in the often intricate chain of mutual adaptive adjustment between a plant species and its animal vectors of pollen or seed. The interdependence is probably most obvious in tropical biota, where the specificity of insects, birds and bats as pollinators is often very high. It will be rare indeed that all the circumstances of the natural environment can be simulated in the conditions of a cultivated collection, and so, notwithstanding Darwin's remarkable detective work on pollination mechanisms in tropical species using largely material from the Kew collections, the contributions botanic garden research can make in this context are likely to be limited.

The situation is quite different where it is the biology of the individual plant that is under study. In the majority of cases such investigations can be pursued effectively using plants in cultivation and often, particularly where laboratory back-up is required, there is no sensible alternative. Aspects of reproductive physiology open to study in cultivated collections include the following:

Flowering

    Relation to:

        Age

        Season and local variation in weather

        Standard botanic garden environments, including
            those under glass;

    Effects of:

        Photoperiod and light intensity and quality

        Temperature and thermoperiod

        Water and mineral nutrition;

Breeding systems

      Flower characteristics: monoecism, dioecism
         and other polymorphisms; dichogamy

      Pollination mechanisms

      Incompatibility systems

      Pollen fertility under different environmental
         regimes

      Pollen tube growth and fertilization;

Seed biology

      Seed set in relation to season and local weather
         variations, and in standard botanic garden
         environments

      Seed viability and storage

      Conditions for germination and seedling establish-
         ment.

The foregoing list subsumes everything from simple observa-
tional work to topics demanding sophisticated research. Where,
then, must one start? We may be assured, first, that for some
disappearing floras practically every solid fragment of information
is valuable, and it should not be concluded that because a full
study of a particular species cannot be carried out it is not
worth doing anything at all. The resources needed to conduct
exhaustive studies are vast, and one might recall that we have
nowhere near full understanding even for some of our most important
crop plants. For wild species, most urgently for those in threatened communities, what is needed is a start in the progressive
accumulation of useful information.

But certain conditions must be established before such
information is likely to be useful in interpreting behaviour in
natural populations or in dealing with the real problems of
management in the field. The requirements for satisfactory research
material include the following:

      Taxonomic authenticity

      Accurate provenance information

      Adequately representative sample size

      Well-grown, disease-free stocks.

No doubt the requirement for taxonomic authentication of
experimental plants will be accepted wholeheartedly by all members
of the conference, but the need has hardly yet been considered by
many plant developmental physiologists. One has only to glance at
current plant physiology journals to appreciate this: sloppy
nomenclatural usage, adherence to vernacular names and use of
stocks of unknown origin and doubtful parentage are commonplace
abuses, and this by workers who would be appalled at the idea of
using contaminated or doubtfully labelled chemicals or inaccurately
calibrated instruments. In the present context, results from
imperfectly identified material would not simply be useless, but
positively harmful to the extent that they could mislead potential
users. The value of voucher specimens deposited in accessible
herbarium collections will be obvious enough.

The indispensability of good provenance information can hardly
be overstressed. Plant species are invariably ecologically
heterogeneous to a greater or lesser extent, with races adapted to
local microclimate or soils. Periodicity in flowering and fruiting
is often related to length of growing season and regulated through
photoperiod, and this type of adaptation is reflected in clinal
variation. The same is often true of temperature adaptation.
These aspects of behaviour can be studied using population samples
in cultivation in gardens, glasshouses or growth chambers, but
the observations will only have ecological meaning if the origins
of the experimental plants are known. I almost feel an apology
is needed for making so elementary a point, but it is certainly
not uncommon for elaborate equipment to be deployed nowadays to
study flowering behaviour of plants grown from seeds picked out of
a seed catalogue, when a little forethought would have permitted
the use of authenticated, localised natural source material and
considerably enhanced the value of the results.

The problems of sampling natural gene pools are adequately
considered in the papers of Professors Hawkes and Esser, and I
would emphasize here only two points. A major goal of ecophysiological research is to explore the tolerance range of species,
and in respect to reproductive physiology one aim must surely be
to map out the ecological "envelope" within which survival and
regeneration is possible. This suggests that wherever possible
evidence should be sought from samples which, on the one hand,
give adequate cover of the variation within local populations and,
on the other hand, offer a reasonable basis for establishing the
total species amplitude, embracing so far as can be arranged the
extremes of the natural ecological range. There are precedents
for studies of this kind for various tree species, to which I
shall shortly refer. The second point concerns the importance of
acquiring a sufficient range of material to embrace an adequate
representation of floral morphological types, incompatibility
genotypes and other variants affecting the breeding system. This

may require deliberately biased sampling programmes, and in some cases repetitive sampling.

For so many distinguished horticulturists it is barely necessary to emphasize the desirability of conducting physiological research on well-grown stocks. Yet, once again, the fact is that experimenters are only too often prepared to accept second best. I have myself seen plants growing in expensive phytotrons enmeshed in red spiders and crawling with aphids - not really living, but simply dying very slowly - but being used nevertheless to generate data. Further comment is superfluous.

Reverting to my earlier listing of the aspects of reproductive physiology open to investigation in living collections, I would like to say something about the assembling of simple phenological data. I have heard the value of this questioned, but there is really no doubt that such information can become important, particularly when the results from different institutions in diverse climates can be brought together and analysed. This has already been shown by work carried out by forestry organisations throughout the world, where performance data have been accumulated for samples of various provenances in different sites and successfully used in determining planting policies. Without the incentives given by the practical needs of forestry, those managing botanic gardens and other general collections have not thought it necessary to assemble data and analyse them in the same way. Yet comparative surveys involving groups of related taxa and taking into account performance parameters like seed set and viability would undoubtedly throw up valuable information. Planned co-operative projects for special groups claiming attention now because of their decline in the wild could hardly fail to provide evidence that could be used in interpreting behaviour in the field, and perhaps in guiding management policy for regeneration. I encountered an example this year in China, where the performance of bamboo species in cultivation in different sites was being investigated with the aim of acquiring information about development, including flowering and seed set, that might be applied in promoting the survival and regeneration of natural bamboo forests now languishing because of earlier misuse.

Many of the possible work programmes listed earlier require laboratory facilities, and some indeed, would demand elaborate equipment, including controlled-environment installations. Obviously the number of institutions in a position to undertake such work is limited. We have to act now with the world as it is, and this means facing the fact that most of the expertise, most of the competent institutions and most of the resources available for immediate application are in the temperate zone. Yet, as many of the other papers have stressed and most notably the contribution of Dr Raven, the urgent problems are in the tropics. Temperate-

zone institutions must therefore be brought to pay more attention
to the problems of the tropical flora, and this is one justification
for maintaining and building up glasshouse collections as a research
resource. A commitment for this purpose is not necessarily an open-
ended one, since the target is the acquisition of information, and
as Dr Poppendieck notes for the Hamburg <u>Mesembryanthemum</u> collection,
it may be easier to justify the expense involved for a project with
definite aims and a specified time scale.

Ultimately there must be a drive to build up competence <u>in</u>
the tropics, where the problems are on the doorstep. In the meantime, however, it is surely a matter for concern that temperate-
zone botanists are so reluctant to turn their attention to
tropical flora. I sometimes feel despair when I see grant applications proposing more and more intensive work on the smaller and
smaller problems of European and North American flora, whether
from physiologists, ecologists, or indeed taxonomists. Somehow we
have to attract more work towards the tropics, and this means
better treatments of tropical vegetation in university courses.
A reversal of the trend to edge out of modern intensive degree
syllabi all references to the major vegetational types of the hot
countries of the world and any consideration whatever of the
challenging problems - physiological, biochemical as well as
taxonomic and ecological - they offer is surely overdue.

In conclusion I turn to the dissemination of results obtained
from the study of the developmental physiology of threatened and
diminishing species. Obviously observations made in any one centre
are likely to be of little value unless they are co-ordinated with
those from others and the information made widely available to all
for whom they have significance. There is a practical difficulty,
however. The research publication media available are necessarily
geared to handling papers dealing with new findings and new
principles. The results of much of the work suggested will not
necessarily meet these requirements. They might, for example,
take the form of detailed performance data relating to groups of
species, or comparing different genotypes or provenances. They
would not necessarily expose new principles, nor relate to new
phenomena, nor yet involve the application of novel techniques.
They would not - and could not - command wide readership, important
although they may be for the actual users. This situation is not
unique, for there are many branches of science and technology where
large amounts of data of a fairly routine nature are generated.
The solution in the past has been the publication of hard-copy
handbooks, tables and the like. In biology, an example is the
Handbook of Biological Data compiled by Spector and Spector, which
included information about matters such as the photoperiodic
responses of plants. Another is provided by the chromosome catalogues now produced in Europe, USSR and USA. And reference to
these reminds us that routine data can feature largely in floras,

horticultural works, and forestry manuals and reports. But today
with the high cost and delays associated with hard-copy publication
it is improbable that orthodox media could fulfil the need for an
economical, speedy and accurate means of disseminating detailed
data of the kind we are considering. There are, however, alternative models that we might now contemplate. National networks
exist in certain countries for the distribution of detailed
environmental data, and progress is being made in the establishment
of an international environmental data referral system. We can
perhaps envisage the progressive building up of comparable systems
of information exchange for plant performance data, perhaps
initially in a series of bilateral contacts, but thereafter pro-
gressing to wider networks. For the great bulk of the information
on detailed matters like temperature and photoperiodic responses
or seed storage and germination requirements, open publication may
not be necessary at any stage, provided only that every institute
or organisation with a legitimate interest can gain access to the
data pool.

Unification in documentation and data handling methods is
clearly a prerequisite for any such system, and some aspects of
this are considered in papers from the section on documentation.
UNEP may provide a possible agency for handling the work, possibly
even as an extension of proposed environmental data systems; after
all the ultimate aim, conservation of the plant kingdom, lies
within its remit.

AUTECOLOGY

C. D. K. Cook

Botanic Garden, University of Zürich

Switzerland

Autecology is the study of the interrelationships between the individual and its environment. The scope of the subject is presented by Daubenmire (1959). The environment may be divided into climatic, edaphic and biotic categories. Each category can be subdivided into numerous factors. The interactions of the various factors are almost uncountable. Autecology is therefore an all-embracing subject which no one person can be expected to master. At the same time, hardly a botanist or gardener can not claim to do some autecological work. Research has, or should have, a motive. For this meeting our motive is conservation. In many ways I prefer to use the term "management" because conservation of threatened plants usually requires management of its environment or habitat. The scope of the subject forces me to select only a few topics.

It is a naïve thought of many biologists that one can learn the environmental limitations or ecological amplitude of a plant in a growth cabinet, phytotron or controlled environment chamber. For conservation-orientated research an enormous energy output is required and as one can vary only a few factors at a time it is very time consuming. Another objection to this kind of approach is that one usually finds that the limits of tolerance to specific environmental factors, such as temperature, light intensity, humidity, etc. are much greater under experimental conditions than those found in nature. Particular phases in the life history of a plant are often more specialised than others. For example, Schneller (1975) has demonstrated how the distribution of the sporophytes of different Dryopteris species is limited by the environmental requirements of their gametophytes. The sporophytes have a wide ecological amplitude while the gametophytes of the different species each have separate and rather specialised requirements.

A further objection to growth cabinet work is that it is very
difficult to control many biotic factors. For example,
Streptocarpus wendlandii Sprenger, is confined to a relatively small
forest in Natal (Hilliard and Burtt, 1971). In the glasshouses at
the botanic garden at Zürich it is almost a pest, seedlings come
up all over the place. Its manifested ecological amplitude in
Zürich is certainly greater than that in its native forest.
Perhaps, a factor limiting its distribution in nature may be no
more than the lack of a long range dispersal mechanism. In general,
the botanic garden has many advantages over the growth cabinet for
determining which factors are limiting the distribution of a
particular plant species.

Whether for teaching or display the majority of botanic
gardens cultivate a wide variety of different plant. In order to
do this a number of different biotopes must be created including
warm and cool glasshouses, peat beds, bogs, pools, sand dunes, etc.
When new and ecologically unknown plants are brought into the
garden they are normally planted in several different places. This
is really no more than autecological research. Frequently one
finds that a plant grows best under the "wrong" conditions. Which
garden grows its plants under the "right" conditions? Knowledge
gained from this kind of experience is very valuable in conservation

Transplanting in different environments is a standard autecological procedure. From the conservation point of view this
method has some dangers, particularly in the introduction of exotics
The introduction of the serious pest Salvinia molesta D.S. Mitchell
in the Old World can be directly attributed to botanic gardens
(Cook and Gut, 1971). It is the policy at Zürich not to supply
American gardeners and aquarists with potentially noxious Old World
aquatic plants. The flora of Europe is not saturated with species
and thus immune from plant invasions. For example, the North
American species Elodea nuttallii (Planchon) St. John was reported
in 1941 from Holland but in the last few years has become established in Germany, Switzerland and England. Two years ago plants
were cultivated in the botanic garden at Zürich for the purpose of
identification, but E. nuttallii has spread in the gardens and is
difficult to remove. If other gardens want to cultivate this
species I would recommend that they take care to keep it in aquarium
tanks. It is not certain that E. nuttallii will displace rare na
natural species in Europe but the possibility exists.

An important function of botanic gardens is to supply material
for research. It is very important to know the origin of the
supplied material. This is particularly so as different races
often have different environmental preferences. Most gardens
cultivate common local wild plants and I would like to make the plea

that when possible the cultivated material is locally collected and properly documented.   Among wetland plants, how many gardens know the origin of their Phragmites, Typha, Sparganium, Alisma, Nymphaea, Lythrum, etc.?

I would now like to discuss some autecological work on some wetland species that are threatened in Europe.   Ranunculus ololeucos Lloyd and R. tripartitus DC. are amphibious plants with a distinctly atlantic distribution.   They are very vulnerable to environmental changes and have recently died out in many localities.   In botanic gardens with an atlantic climate they are easily cultivated in shallow water.   In Central Europe the change from winter to summer seems to be too abrupt; they can be cultivated in glasshouses but this requires a lot of effort.   It is much better when gardens in the right climatic zone undertake the responsibility for their cultivation.   It is then possible to cultivate large populations with the minimum of effort.   With these two species no garden should grow them together as they hybridize and produce a fertile Fl. hybrid.   International coordination in such a case is very important.

Salvinia natans (L.) All., a free-floating fern is, like the Ranunculus, very vulnerable and has become very rare in Central Europe.   But, unlike the Ranunculus, S. natans is notoriously difficult to cultivate (my experience in many gardens is that plants labelled S. natans are usually S. minima Baker, S. rotundifolia Willd. or even ocassionally S. molesta Mitchell).   Our studies have confirmed that S. natans is an annual, a few plants can be overwintered in a glasshouse but they are usually rather weak.   In a more or less stable aquatic environment S. natans makes very few sporocarps or only microsporocarps.   However, if the water level is dropped the stranded plants make masses of micro- and macro-sporocarps.   The plants die and the spores germinate the next spring.   Incidentally, all textbooks illustrating the Salvinia prothallium have it drawn upside down!

For success Salvinia natans needs a habitat that is flooded in winter and spring, but dries out slowly in summer.   Such habitats have become rare in nature but are even rarer in botanic gardens. Many threatened and endangered species in the genera Damasonium, Elatine, Eryngium, Limosella, Marsilea, Pilularia and Thorella are adapted to live in such vernal pools.   This kind of habitat is difficult to manage but there is a good case for attempting to create it in botanic gardens.

As a final example I will take Aldrovanda vesiculosa L.   It is a spectacular carnivorous plant that traps small animals underwater. The trap is in principle like that of the Venus Fly Trap (Dionaea muscipula Ellis ex L.).   In western Europe it is now apparently confined to one pool by a building site near Zürich and one locality

in South Germany.   In southern Africa it is quite common but this
has not prevented us from undertaking a rescue operation in Europe.
We have found that it tolerates shade and if there is not too much
competition from algae and aquatic vascular plants it grows well in
a wide range of aquatic habitats as long as it is kept away from
fish, terrapins and birds.   As a meat carrying plant it is very
attractive to both herbivorous and carnivorous animals.   In Europe
the increase in amenity value of open water has led to an increase
in the population density of fish and water birds, particularly
Mallard and Swan.   This in turn has led to the decrease in certain
plant species.   Around Zürich reed communities can be established
on the edges of lakes and rivers only when a wire-netting cage is
erected first.   Cages for plants are not attractive but they are
occasionally necessary.

There are still many mysteries concerning the ecological whims
of plants.   For example, at Zürich we are unable to keep Hottonia
palustris L. alive.   From our seed list one of the most requested
plants is Wolffia arrhiza (L.) Horkel ex Wimm.   Going through the
records I see we have almost the same customers each year.   Dozens
of botanic gardens are killing our Wolffia each year.   With
conservation-orientated cooperation between botanic gardens it is
important to learn of failures as well as successes.

REFERENCES

COOK, C.D.K. and GUT, B.J. (1971). Salvinia in the State of Kerala,
     India.   Pest Articles & News Summaries 17(4): 438-447.

DAUBENMIRE, R.F. (1959). Plants and environment: A textbook of
     plant autecology.   ed. 2, John Wiley, New York, Chapman Hall,
     London.

HILLIARD, O.M. and BURTT, B.L. (1971). Streptocarpus.   University
     of Natal Press.

SCHNELLER, J. (1975). Untersuchungen an einheimischen Farnen.
     3 Teil. Oecologie. Berichte Schweizerischen Botanischen
     Gesellschaft 84: (in press).

Adjuncts to
Conservation-Orientated
Collections

SEED BANKS

Walter Hondelmann

Agricultural Research Centre

Braunschweig, Federal German Republic

During recent years the occurrence of genetic erosion in the centres of diversity of cultivated plant species, as Professor Hawkes has pointed out in this volume, can be observed in an ever increasing and in part dangerous amount. On the other hand due to the work of plant breeding our crop plants are becoming more and more uniform. In other words, the genetic diversity or heterogeneity is decreasing to such an extent that genetic variability of our crop plants is in great danger. An impressive example (which also illuminates the agricultural significance of this phenomenon) is the fall of the maize yield in the USA in 1970 caused by the attack of Helminthosporium maydis (Southern corn blight). The overall average drop in yield was 15% and losses of up to 50% were recorded in individual areas. The almost exclusive use of the so-called "Texas Cytoplasm" in the hybrids gave rise to a degree of uniformity which made an epidemic of the disease possible.

A better understanding of these three phenomena during recent times has led to worldwide action for collecting, evaluating, documenting and conserving gene material of cultivated plant species and their wild relatives in order to maintain a broad genetic base for further crop improvement. Therefore it is vital to collect and maintain not only a single genotype or even a group of genotypes (ecotypes, etc.) but to work on the population level whenever possible.

The fact that gene banks, as we call the seed banks in general, are given first priority for preservation purposes depends on the premise that the establishment of natural genetic reserve areas for cultivated plant species as well as for their wild relatives and primitive forms is a rather unrealistic task.

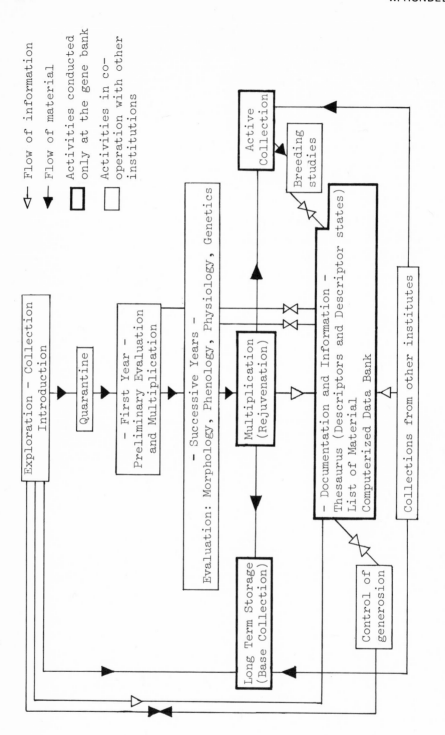

Figure 1. Organisation of gene bank activities (modified from Bari Germplasm Laboratory)

The following activities are typical for a gene bank:

1. Exploration, collection and processing of gene material.

2. Identification, preliminary evaluation and multiplication of the collected material.

3. Initial planting of introduced material in accordance with plant quarantine rules.

4. Exchange and distribution of material with other institutions including private plant breeders.

5. Maintenance and storage of seed (and vegetative material) for short-term and especially long-term storage.

6. Documentation and exchange of information with other institutions.

7. Rejuvenation of gene material if necessary.

8. Organization of training programmes for gene bank personnel.

9. Identification of genetic reserve areas; organization of their exploration.

To put this into diagramatic form to show the organization with existing connections and interactions, etc., see Figure 1.

Having in mind that this paper should be an adjunct to conservation-orientated collections I shall include the following problems:

1. Long term storage for base collections.

2. Genetic integrity of stored seed.

3. Rejuvenation of gene material.

4. Quarantine problems.

5. Documentation and exchange of information.

1. Long Term Storage for Base Collections

If the term "gene bank" is used in the correct context the long term storage of seed is automatically included. The problems of storing seeds over many years has been discussed frequently during recent years. From some experiments and on a broader scale

from practical experience it could be shown that this is no problem
for "normal", ie. non-recalcitrant seeds like those of cereals,
some vegetables, fodder crops, and others, provided that the pre-
treatment of the seed and the storage conditions follow some simple
but effective rules. After cleaning the seed it has to be dried
back to 5%-6% moisture content without affecting viability. To
reach this level some equipment is necessary: either to use warm
air up to 30°-40°C with accelerated ventilation or to use pre-dried
air, for instance employing silica-gel. The last method is to be
preferred in climates with a high relative humidity.

Fatty and oily seeds need an even lower moisture content
(approximately 4%). There are many so-called recalcitrant seeds
within tropical and subtropical fruit species - coffee and cocoa
for instance - the treatment and the storage conditions of which
are not well known.

Before the process of drying is carried out the initial
viability and moisture content has to be determined following the
prescription of the International Seed Testing Association (ISTA).
At this stage with some species one problem might arise: low
viability caused by seed dormancy. Therefore it is necessary to
remove dormancy to avoid confusion. For this some standard methods
are available, for example, to treat the seeds to a low temperature
for some days or to carry out the test in 10^{-3} molar gibberellic
acid.

For use within the network of gene banks FAO recommends the
following conditions for long term storage:

a. Storage at -18°C or below in sealed containers (cans or
   glass jars are better than plastic material) at a moisture
   content of seeds of 5 $\pm$ 1%.

   These conditions are referred to as 'Preferred Standards'.

b. Storage either at 5°C or less in airtight containers at a
   seed moisture content of 5-7% or storage at 5°C or less in
   unsealed containers with controlled atmosphere of not more
   than 20% RH.

   These are referred to as 'Acceptable Standards'.

Recent demonstration at the Russian Gene Bank at Krasnodar
clearly showed the maintenance of a high degree of viability of
cereal seed (with the exception of rye and sorghum) which were kept
with a normal moisture content at a temperature of 4°-7°C for
twenty-four years. According to the estimates of Roberts (1975)
wheat for instance must be regenerated after about 390 years when
using the Preferred Standards (Figure 2).

Figure 2. Estimated Expected Regeneration Intervals
under Various Systems of Storage (from
Roberts 1975)

Organisation	Temper-ature	Moisture content	Estimated regeneration interval, years (time taken for viability to drop by 5% (1)		
			Triticum aestivum	Hordeum(2) distichon	Vicia faba
National Seed Storage Labor-atory, Fort Collins, U.S.A.	4°C (3)	32% R.H. (ie., about 9% mois-ture content for non-oily seeds; about 4%-6% for oily seeds)	9	109	20
Plant Genetics Resources Centre, Izmir, Turkey	0°C	6% moisture content for cereals; 7%-8% for oily seeds	30	710	90
National Seed Storage Labor-atory, Hiratsuka, Japan	-10°C (4)	4%-6% moisture content	120	5,900	440
Vavilov Insti-tute, U.S.S.R.	4.5°C and -10°C	2%-8% depending on crop	Similar to National Seed Storage Laboratory, Japan, when using lower tempera-ture		
Conditions de-signated here as 'Preferred Standards'	-20°C	5% moisture content	390	33,500	1,600
Conditions de-signated here as 'Acceptable Standards'	5°C	7% moisture content	14	220	40

(See p. 218 for footnotes.)

Figure 2, continued.

(1)   Calculated from Basic Viability Equations using constants
      suggested by Roberts.

(2)   Values for <u>Hordeum</u> obtained from experiments on cv 'Proctor',
      which may have an exceptionally long viability compared with
      other barley cultivars.

(3)   Three rooms are also available at -12°C.

(4)   Seed for distribution is held at -1°C.

        For reasons of economics a temperature of -18°C to -20°C should
be employed. Beyond this level, significant increase of costs has
to be taken into account. The trend many gene banks are following,
is to store the seeds at -18°C. Temporary fluctuations of temperature seem to be of no significant influence for seed quality.

        As to the size of accessions there is no general agreement.
From the genetic point of view it depends at first on the breeding
system of the individual crop as shown by Professor Esser in his
paper and also on the heterogenity of the sample. That is to say
that cultivars differ in this respect from populations or primitive
races and also that there is a difference between self- and cross-
pollinating crops as already shown. Obviously there are no experimental results, so that this problem has to be from practical
experience and from known aspects of gene bank management.

        From these considerations it can be stated that sample size
should be at least 10,000-12,000 seeds for cereals in order to meet
all requirements such as to have seed for germination tests, provide breeders' demands or supply Active Collections, etc. For
practical reasons it is recommended that the sample be split into
two parts, one of which is kept for long term storage, the other
one for the aforementioned purposes.

        Within the world network FAO has set up two categories of gene
banks as a result of questionnaires sent out to find the types and
capacities of conservation facilities available.

        The first group comprises the so-called Main Conservation
Centres consisting of sixteen gene banks most of which have adequate
space available, follow up the 'Preferred Standards' and are willing
to store duplicates from other gene banks. The second group, the
so-called 'Associated Conservation Centres', consists of about
fifteen other gene banks, which more or less follow the 'Acceptable
Standards'.

2. Genetic Integrity of Stored Seed

If long term storage is considered to be successful the
maintenance of genetic integrity is a prerequisite. As Roberts
(1973) has shown, loss of seed viability during the storage period
leads to a recognizable amount of chromosomal damage and genetic
mutation. The same author calculated for barley that a viability
loss of 50% shows the same effect as a treatment of fresh seed with
10,00 r of X-rays.

The number of mutations was reduced to about half if viability
was reduced to 80% germination. There are some other reports that
genetic damage could occur if the moisture content of the seed is
lower than 5%, but it has to be pointed out that there are only a
very few (if any) quantitative experiments to support the importance
of genetically caused changes.

On the other hand if seed viability has been maintained at a
high level no significant genetic alteration could be observed.
Therefore the genetic integrity of the seed remains unaltered under
suitable storage conditions like those provided under the 'Preferred
Standards'. These will also help to overcome differential survival
capacities of different genotypes in storage. In this connection an
important question arises: If for instance a user of a gene bank
sample, which in general consists of fifty to one hundred grains,
observes one or two abnormal individuals among the growing plants
(this has to be considered as a very exceptional event), what does
this mean? In most cases the user would discard these plants not
recognizing that they are mutants but rather believing that there
is something wrong with the sample. Therefore from the practical
point of view the occurrence of mutations or of genetically damaged
material must not be over-estimated in respect of their importance.

3. Rejuvenation of Gene Material

There is a rather close connection between genetic integrity
and rejuvenation, because genetic integrity could be strongly influenced by the work of rejuvenation (which sometimes is referred
to as regeneration or generally as multiplication), in order to
produce fresh stock. The interval of rejuvenation depends on
factors like decrease of viability, decrease of sample size caused
by seed supply, germination tests, etc. From the foregoing remarks
on seed viability it follows that rejuvenation should be carried
out if any loss of viability is observed. There are of course some
additional problems involved:

(i) When seed samples have to be rejuvenated, which were
collected in a region with different environmental
conditions, these plants are subject to selection. The

more cycles of rejuvenation are carried out, the more changes in the genetic composition of the sample will occur. After three or four cycles selection sometimes has been of such influence that the original composition cannot be recognized. The way out of the difficulty would be to do rejuvenation work only under those environmental conditions in which the samples were collected. This would require a very advanced stage of international co-operation, that might never be reached. Therefore rejuvenation should be the exception.

(ii) A general problem among cross-pollination crops is that of outcrossing with other samples. Therefore it is necessary to have isolation cabins or something similar. If not then natural isolating conditions must be carefully chosen. We ourselves are studying these problems using marker pollen grains with rye in order to get quantitative measurements in relation to distances and some meteorological data.

(iii) From population genetic studies it is well known that small populations are drifting towards fixation with respect to certain alleles. Therefore after some generations homozygosity could be reached to a high degree. But also with this problem it has to be pointed out that we are waiting for further quantitative experiments which will help to clarify the special situation with regard to gene bank material.

4. Quarantine Problems

If we have seed exchange between two countries or the introduction of exotic stocks to one country there is of course the danger of introducing pests and diseases together with gene material which was not known before. Foreign germplasm as a resource for new phytosanitary problems is a frequently discussed question.

Therefore strict observance of plant quarantine regulations is necessary. In general there is a legal basis for plant quarantine rules and there is the International Plant Protection Convention. It is therefore not necessary to go into further details here. But it should be stated that if a country does not have these regulations but requires gene bank activities to be carried out adequate facilities for plant quarantine work would have to be prepared.

5. Documentation and Exchange of Information

From Figures 1 and 2 it is clear that necessarily there is a flow of information between the different activities of a single gene bank and moreover between different gene banks within the world network. Furthermore, gene bank samples without any information are useless. The most effective form of mutual information is that which can be electronically displayed and computers for this purpose are now available everywhere.

The most profitable computer involvement in relation to documentation activities is guaranteed by means of managing only the essential part of the information. Such essentials are found in all data that represents the informative content of a document. Usually tables are designed for the transmission of gene bank data. Indeed, tables can be found in gene bank documentation, eg. Index Seminum, seed exchange lists, or illustrating the results of material evaluation.

With the exception of descriptive terms used in a tabulated document the informative data related to the descriptive terms are almost universally understandable, since they are normally used in the form of proper names (varietal designations, etc.) or figures (metric dimensions), ie. in comparison with descriptors they do not need translation into another language. Two technical terms are used in this connection: descriptors and descriptor states. Descriptive terms are called descriptors; the "informative data" by means of which a descriptor becomes more precise are the so-called states of a descriptor. For the better use of computing machines there is the need for condensation of gene bank data, ie. for any data to be given away or for external data to be incorporated into a data bank a standardisation of the format for data exchange is recommended.

We ourselves are using the so-called data processing system TAXIR. The TAXIR system has been recommended as useful for documentation purposes of gene banks. Its adaptation, however, requires computer facilities that are not available everywhere. We decided that the TAXIR data format should be chosen for our purposes since it seemed to us obvious that all gene banks with access to TAXIR adequate computers would make use of this system. Indeed, the development has confirmed this. With this prerequisite we consider our data to be easily adaptable to those of other gene banks that work with TAXIR or to incorporate the TAXIR processed data from other gene banks and our electronic files of data. The data processing system used at Braunschweig is designed for a Siemens computer. Only a few programming arrangements were needed in order to adapt our TAXIR-formulated data to a data storage and retrieval system that is used commercially in our country.

TAXIR in itself is a data storage and retrieval system, it has been continuously developed further on to EXIR and EXIS ie. Executive Information System, by the Boulder Group (Professor Rogers).

With respect to further computer requirements the layout of magnetic tapes has already been standardized internationally. We therefore suggest IBM-compatible tapes with nine tracks and 800 Bpi to be used for international data exchange.

For the economic use of computers brief information is necessary, ie. codes and abbreviations are common. But from general experience the conclusion was reached that descriptors should never be coded. Descriptor states are very well suited for the employment of codes. Suggested German and English descriptors and partly Spanish descriptors in combination with either coded or uncoded descriptor states have been compiled in a "Thesaurus for the International Standardization of Gene Bank Documentation" (Seidewitz, 1973). This Thesaurus has to be understood as a dictionary for international co-operation. Descriptors though available in different languages are computer technically considered as synonyms. Corresponding arrangements have been made in our gene bank documentation systems that a descriptor in a foreign language can be retrieved by means of a German descriptor.

For the worldwide use of the descriptors and descriptor states there are rules which are compiled in the Supplement. These rules were agreed upon for wheat during the Wheat Genetic Symposium of IBPGR in Leningrad this year following the suggestions of Mr Seidewitz of our gene bank (Rogers and Hersh, 1975). It is to be expected that other crops will follow in the near future.

Finally I would like to point out that from a purely scientific point of view some of the points mentioned above have not been examined or clarified satisfactorily. However, for the great task of reducing the genetic vulnerability of our crop plants, the establishment of seed banks is the only practical way of preservation which can be followed everywhere. It also must be stressed that even a well-operated world network can only be attained by setting up priorities in respect to crops and regions.

REFERENCES

FAO. (1974). <u>Meeting of Panel of Experts on Plant Exploration and Introduction</u>. Rome.

NASC. (1972). <u>Genetic vulnerability of major crops</u>. Washington D.C.

ROBERTS, E.H. (1973). Loss of seed viability: chromosomal and genetical aspects. <u>Seed Science and Technology</u> 1: 515-527.

ROBERTS, E.H. (1973). Predicting the storage life of seeds. <u>Seed Science and Technology</u> 1: 499-514.

ROBERTS, E.H. (1975). Storage of wheat seed for genetic conservation. <u>In</u> IBPGR <u>Symposium on wheat genetic resources</u>. Leningrad.

ROGERS, D.J. and HERSH, G.N. (1975). Genetic resources communication information documentation system GR/CIDS. <u>In</u> IBPGR <u>Symposium on wheat genetic resources</u>. Leningrad.

SEIDEWITZ, L. (1973). <u>Thesaurus for the International Standardization of Gene Bank Documentation</u>, Parts I-IV. Braunschweig.

SUPPLEMENT
[from Rogers and Hersh (1975)]

Descriptor Observation/Measurement Scales:

(i) Whenever possible descriptor observations/measurements will be made using the MKS system (meter-kilogram-second), or scales computed and derived from this system - especially for continuously expressed character states.

For directly observable/measurable discreetly expressed descriptor states standard reference scales will be suggested (colour, shape, etc.). The descriptor name will include the scale name used: eg. plant height in cm, etc.

(ii) Where relative/qualitative observations are recorded these are taken for communication purposes only and generally cannot be used for further processing of patterns or for inference. Here an 0-9 scale is suggested where '0' is reserved for the expression of a zero observation <u>made</u> on a descriptor, and a blank (' ') is used to indicate that no observation has been made on a descriptor (unknown). 1-9 are used with '1' being the minimum expression of the descriptor and '9' the maximum expression.

It is suggested that for relative/qualitative descriptors the
range from which the relative scale is derived be provided
for each state as well as the qualitative interpretation for
each state, eg. Plant height 1 = very short (1-50 cm).

(iii) For pest reactions the following scale is suggested and may
be used in combination with the 1-9 scale; HSTRI (Hyper-
sensitive, Susceptible, Tolerant, Resistant, Immune).

(iv) Descriptors may be chosen for use by each centre as needed
but should be defined rigorously and submitted for registration and publication in the International Reference Manual;
however, minimum descriptors should be used to facilitate
international genetic resources work. Minimum lists should
be established for each genetic resources function.

(v) Data transmission among centres should be according to the
following rules:

Data can be transmitted in machine-readable or non-machine-
readable form. The following is proposed:

If machine-readable:

a. 80-column standard cards IBM punch code equivalent or high
quality magnetic tape should be used. The punch character
set codes used should be indicated (a copy made and sent
with the data).

b. Magnetic tape should be 9-track IBM compatible, external
BCD (Note: punch conversion by the sender if tape, by the
receiver if on punch cards), 800-BPI, 80-character records,
not labelled with separate files for data and descriptive
information. The tape should be accompanied by a memorandum
stating tape serial number, place of origin, computer used,
job control language used in processing, number and names
of files contained. A listing of the data and the recording
template should accompany the tape or cards.

If non-machine-readable:

a. Either columnar pads (80 or 120 column) or photocopies with
field definitions used on the first sheet of all original
records should be sent, accompanied by the recording
definitions. Letters should be crossed, not numerals.

b. Transmission should be by mail, diplomatic or FAO pouch.
Data should be receipted immediately. Assistance should
be provided from the GR/CIDS team for planning this aspect,
if required.

THE ROLE OF SEED LISTS IN BOTANIC GARDENS TODAY

V. H. Heywood

University of Reading

England

The Seed List or <u>Index Seminum</u> is one of the most curious types of publication in the history of botany. Indeed it almost constitutes a diagnostic character of Botanic Gardens. Like many other aspects of Botanic Gardens which have been considered at this symposium, its role has changed considerably over the past 100-150 years although its new role in the modern botanic garden is still not quite resolved and many seed lists are still basically similar in content to those published fifty years ago.

Seed lists today still afford an almost bewildering diversity of size, format, style, content and degree of luxury of presentation. They range from a single cyclostyled sheet to a glossy publicity-type volume with one or more illustrations, even in colour. The economics of the printing of the latter must be highly suspect in view of present day costs, and it must be said that luxury of presentation is seldom correlated with the value of the contents! But basically they are all trying to continue a long-standing function - the offering in exchange to other botanic gardens of seeds or other propagules available for distribution. The total number of seed lists produced is difficult to estimate accurately but is of the order of 600-700. Each year a number of new ones comes to our attention in the annual seed list season - they flow in rather like Christmas cards, and like Christmas cards the season seems to extend year by year.

Few detailed considerations have been made of seed lists, although in recent years a number of recommendations have been published about possible standardization of their format and contents. The International Association of Botanic Gardens has discussed seed lists on more than one occasion and a paper on them,

incorporating suggestions made at the sixteenth International
Horticultural Congress in Belgium in 1962 and in subsequent correspondence, was published by Howard et al. in 1964. I myself published
a paper on some aspects of seed lists and taxonomy at the same time
and later in the year a list of botanic gardens offering seed of
spontaneous plants was published by me in taxon on behalf of the
International Organization of Plant Biosystematists. This list
contained 185 gardens which offered at least some wild-collected
seeds, and since then the number has probably increased to about
250. Suggestions for the use of seed banks as a means of improving
the quality of seed lists were made by Thompson (1970).

        The majority of seed lists, both general and those listing
wild-collected seed, are issued from gardens in the Northern
Hemisphere, reflecting the uneven distribution of botanic gardens
in the world. A curious feature is the relatively small number of
gardens in the USA offering lists and, of course, those issued from
tropical gardens is lamentably small.

        The Seed List seems to have had its origin in the catalogues
published by several European botanic gardens in the nineteenth
century of the holdings of plants they cultivated. The early
botanic gardens were, as Professor Heslop-Harrison has described
them, resource centres. The different gardens vied with each other
in the numbers of species they could introduce from overseas
territories – and special expeditions were often sent out to collect
seeds from various parts of the old and new world, especially of
plants of medicinal or economic importance, as well as decorative
plants, in the seventeenth and eighteenth centuries. It was only
natural that the acquisitions should be publicized in the form of
catalogues. It was implicit in the publication of these catalogues
that material of the taxa listed were available on request in many
cases, following the long-standing tradition of international co-
operation through exchange of material. Seed is, of course, a
particularly convenient vehicle for such exchange.

        Later in the nineteenth century the seed list as we recognize
it today evolved and with it the complex and increasingly expensive
series of operations involved – collection, cleaning, packeting,
storage of the seeds, preparation of the lists, printing, distribution, postage, receipt of requests, meeting these requests,
further postage, receipt of other gardens' lists, request of seeds,
receipt, etc., etc. Many gardens established special sections with
their own staffs to look after these activities. The expense involved does not seem to have been seriously queried in terms of
value for money, at least until recent years: the Seed List was to
many a garden an essential and inevitable feature of its existence.

        As botanic gardens grew in number, size and importance, and
as the range of taxa cultivated increased, running into tens of

thousands, the lists tended in many cases to become overlarge, unselective, inaccurate and repetitious in the sense that they often contained the same material, often received shortly before from some other garden. As several contributors to this symposium have already mentioned, a more or less standard set of taxa was built up in many botanic gardens, often at the expense of local native taxa which were largely ignored. The Seed Lists reflected this and the same often inaccurately identified seeds were exchanged, cultivated, exchanged and cultivated year after year, often ending up again in the garden which first issued them. The often considerable percentage of taxonomic misidentifications of plants grown in botanic gardens until recently is a direct consequence of the receipt of misidientified seed received through the Index Seminum system. Most readers will be aware, I am sure, of the very high percentage of misidentification in many genera. Over many years I have worked on _Digitalis_ and calculated the percentage of incorrectly identified seed received as 70%-80% despite the small number of taxa in the genus and the ease of their identification. Quite high figures for commercial seeds of cultivars distributed through catalogues have also been reported.

The consequences of this are very serious, especially as until the post-war years little attention seemed to be paid to this question by cytologists, geneticists, plant breeders, etc. As an example, I can cite the classic paper of hybridization in _Digitalis_ by Buxton and Dark (1934) where the fortunate circumstance that they published coloured photographs of the corollas of the taxa they worked on allowed me to ascertain that many of the taxa were wrongly named. The results had naturally been accepted until attention was drawn to this some fifteen years later.

In the matter of early chromosome counts, the consequences of misidentification are again well-known and should need no stressing today. But although cytologists and certainly cytotaxonomists have, by and large, restricted their recent work to wild-collected seed, or plants raised from them, a new problem has arisen with the development of phytochemistry or biochemical systematics where the same lesson has to be learned again – and of course, in reproductive physiology as Professor Heslop-Harrison stressed in his paper. In many such matters we have to educate our own colleagues!

Now while one could say that the onus is on the scientist to ensure the accuracy of the identification of the material he works on (_caveat emptor_!), there is a strong case of botanic gardens not to issue seed in their Lists of Plants whose identity has not been verified. This would have a dramatic effect on some lists and would, even if applied, not entirely relieve the user of the burden of checking. Some gardens have already adopted such a policy or at least indicate which plants in their list have been verified, and some publish a list of corrections. Certainly, it is on reflection curious that serious establishments should publish ostensibly scientific lists which are known to contain a high percentage of errors, even though we all know the problems of manpower, time and expertise needed to avoid this.

There is, of course, a case to be made for only seed of known origin in the wild, or seed immediately derived from plants raised from such seed under controlled conditions, to be offered in seed lists, apart from horticultural cultivars.

In fact, one of the main changes that has taken place in the content of seed lists in recent years has been the increased emphasis on offerings of seeds of wild origin, usually collected by botanists during field excursions, either in their own country or abroad. The pressure for listing such seed has come largely from changes in the practice of taxonomy, especially with the development of biosystematics or experimental taxonomy. Seed lists have now become one of the main sources of seed for such studies and consequently the "audience" has also changed - lists are eagerly sought by taxonomists and other research workers in botanical institutions as well as by horticulturists. Attention therefore needs to be paid to the circulation of such lists - normally they are received by Garden Staff and circulation even within an institution poses problems in that scores of people may wish to check through them. Certainly, they ought to be circulated to all interested institutions, not just to botanic gardens.

As regards the offering of spontaneous/wild seed, a number of suggestions were made in 1961 by the executive of the International Organization of Plant Biosystematists at its Neuchatel meeting (Heywood, 1964) and these have been acted upon by many gardens, although by no means all. The points may be repeated here and further kinds of information could no doubt be added.

1. A distinction should be made between seed collected from wild plants in natural habitats and seed collected in botanic gardens from wild plants raised there.

2. Spontaneous seed should form a special part of the seed list, distinct from the sections giving ornamental and other seed.

3. For each collection there should be given the name, authority, station (country, county, etc.), altitude, habitat (vegetation, soil, if possible).

4. Each collection should be numbered, and seeds ordered by number.

5. These lists should be circulated to all interested institutions, not just to other botanic gardens.

This change of emphasis in the content of seed lists forms part of the trend from quantity to quality.

A further major change in the whole question of seed lists is very recent and follows from studies on seed storage and seed banking and from the whole question of resource and conservation studies. In the light of these developments, consideration should be given to including information in seed lists, in addition to accurate identification already mentioned, and the IOPB recommendations just noted, such as seed viability, condition, conditions for germination, conditions for establishment, seed set, etc. There is an enormous amount still to be learned about such features in most species and it would be very valuable if gardens could indicate what information they have available when listing seeds. It is quite evident that a great deal of information is acquired by botanists in institutions throughout the world about the storage and germination characteristics of taxa as part of their biosystematic/experimental taxonomic studies. Perhaps they could be persuaded to offer this information to the compilers of seed lists in their institutions. This in turn would lead to the need for some collation and centralization of this kind of information, perhaps through the kinds of data banks mentioned by other contributors. A case could also be made for indicating in seed lists data on other aspects of reproductive physiology such as those mentioned elsewhere in this book by Esser and Heslop-Harrison.

We should study whether the seed list in its modified sense with its rather closed circulation is a suitable vehicle for dissemination of this kind of data - it is not the open type of publication which Professor Heslop-Harrison has suggested as not necessarily appropriate for data on reproductive biology but occupies an intermediate position and could well reach the desired scientific public.

Such mechanism would not replace the use of centralized EDP systems such as TAXIR, etc. but it is not realistic in the short or even medium term for us to envisage the majority of botanic gardens becoming associated with such EDP systems. One has to bear in mind that the majority of botanic gardens in the world, even in Europe, are not as well or as securely financed as some of the major ones, such as the Royal Botanic Gardens at Kew and Edinburgh, and the Botanic Gardens of New York, Missouri, Hamburg, etc.

A word ought to be added about the condition in which seed is stored by gardens while awaiting requests. While not all institutions can follow the example of Kew in its latest Seed List which offers material from conditions of controlled storage from its seed bank, it is all too clear that much seed is gathered and stored by gardens in inadequate conditions and has lost its viability even before it is sent out to requesting gardens. As Howard _et al_.

noted in their 1964 paper "the goal of any seed list is to offer viable seed.   Any practice which affects this should be eliminated".

In addition to offering better quality, better documented, better stored seeds, another class of information now needs to be considered.   I refer to the rarity and conservation status of the taxa offered.   It is, I venture to suggest, liable to lead to a misuse of resources to offer seed of relatively rare, endemic species openly in Seed Lists without giving any indication as to rarity.   There are normally no, or very few, restrictions placed on the user of Seed Lists, and consequently what may be a very rare resource, often obtained under conditions of great difficulty and discomfort may be squandered by the casual requester.   Perhaps one ought to make an enquiry into the percentage of packets of seeds requested that are actually used – and I must admit that it is the taxonomist rather than the horticulturist who is the chief offender here!

Just as restrictions have to be built into the utilization of seeds from seed banks, so similar restrictions may have to be placed on the use of seeds from normal seed lists.   Since we are witnessing a gradual merging of catalogues of holdings in seed banks and Seed Lists in the more traditional sense, such a development seems a logical conclusion.

To summarize, seed exchange should become a much more critical and selective activity.   Since it will cost more, especially in terms of scientific staff time, the number of taxa listed will have to be reduced, allowing Gardens to concentrate on their native floras, specialized collections (on a geographical and taxonomic basis) or on exotic floras in which they have a special research interest.   The question of listing and documenting cultivars is another matter which I shall not go into although the problems involved should not be overlooked.   The relative responsibilities of Botanic Gardens and commercial Seed Establishments and indeed of Agronomic Institutes in such matters is a difficult question and one that deserves detailed analysis.

Another area where Botanic Gardens may be able to play a role directly or through their Seed Lists is in the matter of discontinued research projects where stocks of material and seed are built up over a period of years and are liable to be discarded or dispersed on completion of the programme.   The material could be offered through Seed Lists to other possibly interested workers.

Although it is not the primary concern of this paper, a few comments may be made on commercial Seed Lists and Catalogues.   These normally list a large number of cultivars and large numbers are printed for potential customers who are usually restricted to the country where the seed catalogue originates.   Commercial seed catalogues are aimed at selling seed and other related items and are often beautifully produced and printed.   Often cultural instructions are given unlike the situation in Botanic Garden lists where it is assumed (often erroneously) that the recipient will know how to germinate and grow the seed.

A point that should not be overlooked is that commercial seed catalogues are a valuable source of information on the development and spread of cultivars and this is of increasing importance in these days when economic circumstances are causing many firms to restrict the number of cultivars they maintain and offer.

CONCLUSION

I have tried in this paper to draw attention to some of the points that sould be taken into account by Botanic Gardens and relevant staff and by users in the whole Index Seminum operation. It is not intended to be exhaustive and I am sure that many further suggestions could be made.   Perhaps the whole question of Seed Lists ought to be studied by a working party with a view to formulating a comprehensive list of recommendations which could then form the basis of a resolution for a future Congress or meeting of the International Association of Botanic Gardens.

REFERENCES

BUXTON, B.H. and DARK, S.O.S. (1934). Hybrids of Digitalis dubia
     and D. mertonensis with various other species.   Journal of
     Genetics 29: 109-122.

HEYWOOD, V.H. (1964). Some aspects of seed lists and taxonomy.
     Taxon 13: 94-95.

HEYWOOD, V.H. (compiler). (1964). List of botanic gardens offering
     seed of spontaneous plants.   Taxon 13: 137-142.

HOWARD, R.A., GREEN, P.S., BAKER, H.G. and YEO, P.F. (1964).
     Comments on "seed lists".   Taxon 13: 90-94.

THOMPSON, P.A. (1970). Seed banks as a means of improving the
     quality of Seed Lists.   Taxon 19: 59-67.

DISCUSSION SECTION VIII ADJUNCTS TO CONSERVATION-ORIENTATED
COLLECTIONS

Professor J Heslop-Harrison (Kew) commented on the papers by
Professors Hondelmann and Heywood and the mention of a new-style
list based on holdings of a bank. He stated that there was a
component of holdings in botanic gardens which required some
publicity but which might not reach the standards of a main list.
It would be necessary to supplement the seed list with a further
list of material available from gardens, also seeds and other
propagules, on demand.

Dr Walters (Cambridge) said that a seed list is a publicity
organ of a specialist kind. It can be made more useful if
information of interest and value to other botanic gardens can be
inserted in the list. For example, data on annual rainfall is
included in the Cambridge Botanic Garden's seed list.

Professor Heywood (Reading) agreed with Dr Walter's suggestion
and said that such suggestions are listed by the IABG.

Dr Hall (Cape Town) said that seed banks could perform a very
useful bridging operation for holding material until suitable
habitats could be found. He asked whether there was any difficulty
in banking small-seeded plants such as Ericaceae.

Professor Hondelmann (Braunschweig, West Germany) replied that
he had had no experience with Ericaceae, but there had been no
problems with very small vegetable seeds.

Professor Gomez-Campo (Madrid) said that his experience with
Arabidopsis seeds confirmed Professor Hondelmann's statement.

Sir Otto Frankel (Canberra) pointed out that with agricultural
plants a much larger diversity is required by breeders than is
generally necessary for wild plants in maintaining a representative
sample of the population. The population of a plant held as seed
may be changed through ignorance of the natural requirements such
as vernalisation. He also mentioned that the case of plants which
could not be reproduced by seed had not been considered. The
problem had not been solved in agriculture, and was also important
in horticulture.

233

In answer to a suggestion that certain countries did not have or could not afford computerisation of records he pointed out that the cost of centralised computer records need not be high, and this is important for small botanic gardens.

Dr Stearn (British Museum, Natural History) mentioned the dangers of collecting seeds from incorrectly labelled plants and said that the scientific verification of the identity of plants from which seed is gathered is very important.

Mr Shaw (Edinburgh) pointed out the problems caused by open-pollination in collections which might give rise to hybrid seed.

Professor Gomez-Campo (Madrid) added that keeping pollen collections provided a means of retaining genetic variability in a much smaller space even than that required for seeds. Large amounts of pollen could be collected without damaging a natural population, and this may be important when dealing with narrow endemics.

Botanic Gardens
in Relation to Public Education

SECTION IX BOTANIC GARDENS IN RELATION TO PUBLIC EDUCATION

COLLOQUY: EDUCATION IN RELATION TO PLANT COLLECTIONS

Chairman Mr D Henderson, Royal Botanic Garden, Edinburgh

Introductory contributions by:

Mr D Henderson
     Royal Botanic Garden, Edinburgh

Dr V E Corona and Mrs Magdalena Peña de Sousa
     National University of Mexico Botanic Garden

Mr R Hebb
     New York Botanic Garden

Mr Henderson (Edinburgh) stated that botanic gardens deal with research, education and amenity.   Five to ten million people probably visit the gardens of Europe annually, and a challenge to such gardens is to make collections more interesting and intelligible and to compete with zoos and museums with good displays.   A high proportion of visitors had some idea that scientific research was carried out in botanic gardens, although most visited gardens for relaxation and tranquillity.

New methods in labelling, geographical layout, taxonomic arrangement, use of specialised collections, guide books, taped guides (fixed or portable) had to be developed.   A basic problem of education was how to display plants to the public.   School visits could be so arranged as to combine plants in a botanic garden with animals in a zoo showing species from the same geographical regions.

Dr Corona (Mexico City) described the University of Mexico Botanic Garden and said it was founded in 1959 with three objectives: to promote botanical research on Mexican plants, to facilitate teaching at the University, and to encourage botanical education of the general public.

237

The latter two involve close contact with the public and
students.   The collections were labelled and organised in an
ecological manner representing floras of different areas of the
country.

Mention was made of the climatic regions of Mexico, which is
thirty per cent tropical rain-forest and seventy per cent arid, and
also of the creation of satellite gardens by local universities, the
tropical rain-forest station in the state of Vera Cruz, the tropical
west coast station in Jalisco, and another to be established in San
Luis Potosi.   A team of well-trained botanists was being built up
at various centres; the country was ready to act for conservation
but there were economic difficulties.

Mrs de Sousa (Mexico City) said that the Extension Services of
the University of Mexico Botanic Garden hope to encourage a wider
appreciation of the native plants of Mexico.   Films have been made
of the extensive orchid collection from a horticultural and botanical
point of view and courses are held on the morphology and floral biology of the orchids which form a conspicuous part of the Mexican
flora.   Instruction on the horticultural requirements of orchids
has been popular and by encouraging the public to use alternative
growing media the destruction of the natural bark habitats has been
minimised.   Field excursions are also made to show the public
particular vegetation types or plants of special interest in their
natural environment.   Conferences are arranged and radio and
television as well as published material are used to promote the
conservation activities of the Botanic Garden.

Mr Hebb (New York) said that unlike European botanic gardens,
where there is usually a passive involvement of the public, those
in the United States attempt to encourage active involvement.   This
is both necessary and desirable since the gardens rely heavily on
financial support from the public and from private bodies.

The New York Botanic Garden offers 150 instruction courses
catering for up to 60,000 schoolchildren in a year.   The children
are brought to the garden by their teachers and are taken, for
example, on a tour of the woodland where they are taught some aspect
of forest biology.   Children are supplied with garden plots where
they can grow vegetables to take home, practical work which also
serves to encourage interest in plants.   Courses for adults are held
on horticulture, landscape gardening and plant identification.   A
complete two year course in botany leading to a diploma is also
available to the public.

The Cary Arboretum, of about 2,000 acres, allows research on
environmental problems.   The expertise of the staff is often
employed on a contract basis, such as advising Power Authorities on
the laying of power lines with the minimum disturbance to the

countryside and to any threatened species therein.

Membership of the Botanical Garden is offered for a nominal fee. The 10,000 or so members each receive a publication of popular information, reduced rates for courses, and are offered surplus plant material. A small shop is proposed for the Arboretum which will disseminate surplus material of rare and endangered species.

DISCUSSION SECTION IX BOTANIC GARDENS IN RELATION TO
PUBLIC EDUCATION

Professor Cook (Zurich) speaking of the Zurich University
Botanic Garden, said that it is a new garden which is still in
the process of being built up. The support of the local populace
is essential because it is they who vote the money for the garden.
The educational aspect of the garden is decentralised, no comprehensive guide being available. However, when plants of particular
interest are in flower or fruit a descriptive note is placed
beside them, but only so long as the flower or fruit is retained
on the plant. It is felt that permanent placards would disturb
the overall pattern. Taped lectures are available in a darkened
underground chamber from which the tropical aquarium can be viewed
from beneath. Museum cabinets on various botanical themes are
scattered about the garden, for example in the restaurant. Public
lectures are given, also conducted tours of the garden, and so
great is the response that numbers are now reaching the maximum
that can be dealt with.

Mr Apel (Hamburg) pointed out that education is a function of
a botanic garden necessary to gain scientific and public interest.
At Hamburg a central systematic garden has been made in the
arrangement of a phylogenetic tree covering three hectares. The
arrangement, based on Taktajan's scheme demonstrates the evolution
of plants from a single celled alga to the highly developed
orchids. This garden has created a great deal of public interest.
As well as the information given to visitors in guide books and on
plant labels Mr Apel stressed the need to interest the public
directly in other aspects of botany such as food plants, insecti-
vorous plants and in conservation. The education of schoolchildren is most important and at the Hamburg Botanical Garden
teachers are invited for talks prior to bringing their classes to
lectures on various botanical subjects. By such means it is hoped
to influence public opinion and convince them of the botanic
garden's functions including conservation, and thereby gain
governmental support to finance the construction of display houses
and other buildings.

Professor Esser (Bochum) suggested collaboration in writing a
botanic gardens manual which might contain various kinds of
standardised information that could be used in the preparation of
accurate and informative labels, notes on computerisation and other

241

information retrieval systems.   Such a publication would allow
ideas and expertise to be shared throughout the botanical and
horticultural world, and as a ready reference it would save a
considerable amount of time.   The sharing of exhibits was also
suggested, to make them available to a wider public.

Professor Bocher (Copenhagen) mentioned the recent two very
successful Open Days at the Copenhagen Botanical Garden during
which 5,000 members of the public visited the laboratories.
Tours of the glasshouses were arranged and an exhibition was staged.

Dr Y. Heslop-Harrison (Kew) said that it should be appreciated
that Kew and similar botanic gardens which are government-supported
suffer from some disadvantages.   The public cannot be exploited to
help build up the exhibitions side as is done by some non-
government gardens.   At Kew there is a staff of guide lecturers
who conduct parties of both adults and schoolchildren round the
living collections.   A supply of informative leaflets is available
in the various museums and at other points of interest, also quiz
sheets with a range of simple questions for children to answer.
These stimulate them to read the exhibits, think about the subject
and look at the plants in a more careful and constructive way.

Mr Sipkes (Rockanje) mentioned the 'Heemparken' or
'Instructieve Plantsoenen' of which there are about twelve in the
Netherlands.   Their aim is to teach people the value of unpolluted
land and water and how to look after plants and birds.

Sometimes these gardens have a wider function.   In one area
where the sand dunes that acted as sea-defences had been disrupted
by major trench digging he had the opportunity to restore the
former vegetation which consisted of Rosa spinosissima var.
pimpinellifolia and Salix repens var. arenaria.   He was also
pleased to report the success of growing Anthyllis vulneraria
experimentally to improve the soil nitrogen content and allow the
original species to recolonise.

International Co-operation
and Legislation

THE TIME SCALE OF CONCERN

Sir Otto Frankel

CSIRO

Canberra, Australia

Why conservation? Quite a number of evolutionary geneticists believe that man as the predominant organism has the right to use and destroy whatever he cares to because we are the predominant species and the world, living and non-living, is ours. Has there been another species in evolution which cared about other species? So, why conservation?

The reasons given have been economic, scientific, social, educational - is there not an ethical reason? This is a personal question with no general answer. I refer you to an important book by John Passmore, "Man's responsibilities to Nature in the light of Western Civilisation". He discusses man's attitude to nature which started with Noah and continued through the period so strongly influenced culturally by the Bible, through the Greek tradition and ended with the Graeco-Christian tradition which paid no attention to life and made man entirely supreme. This tradition was with us until the end of the eighteenth century. It is quite contrary to the eastern tradition, but our own western tradition was unconcerned with the matters which concern us today. Today we have become the most powerful and most destructive organism, but some of us feel we have responsibility - an evolutionary responsibility. We may be developing an evolutionary ethic.

It is only among people of like mind that we can talk seriously about such matters. Passmore thought that for any generation to give up things which are considered essential was irrational. I feel that it is likely that future generations will develop an evolutionary ethic.

Having said this, I feel that many will be unconvinced, but I

245

believe we should think about the philosophical background and not
only the utilitarian.

In crop conservation the emphasis is mainly on the primitive
cultivars, which were developed over thousands of years and which
evolved with migration etc. since the beginning of agriculture ten
thousand years ago. These were, and some remaining ones are,
reservoirs of genetic diversity which evolved under the impact of
natural selection under cultivation. These were extremely
important in the evolution of the crop varieties by which we now
exist and without which we should starve.

The advanced cultivars are very restricted in their genetic
diversity because of pure breeding between them, because many are
related, and all selected for the same ends - high economic performance. The gene pool of the advanced cultivars is much smaller
than that of the older ones. We are concerned also with the
preservation of some, relatively few, wild cultivars. The primitive
cultivars which have been removed by the advance of modern cultivars
are the main concern of genetic conservation.

In botanic gardens efforts towards conservation of wild species
is the main objective, irrespective of whether or not they are useful to man. FAO and other organisations cover the cultivars of
importance and their wild relatives. Wild species of no particular
economic importance are the focus of botanic gardens. In crops the
genetic diversity is essential for use because the selections one
would make from the primitive cultivars are the things which really
matter. We select for particular disease resistance and particular characteristics of potential or actual use to man. In the
wild species the diversity is merely needed for survival. In
botanic gardens the interest is in survival of the species; in
crops there is substantial financial support for the preservation
of primitive cultivars. The resources of botanic gardens are
limited when compared with economic crop interests and there are
limits of man-power. Conservation is an adjunct to botanic garden
work but it is not the main purpose, as it is in crop work. The
material, interest, use and resources are all different, thus it
is more necessary for botanic gardens to be highly economical in
the use of their resources. As Professor Heslop-Harrison said,
one has to be highly selective in the material to be preserved.

Each garden can have only limited objectives for conservation,
but a combination of botanic gardens can have much wider objectives.
The basic question is - what is the end use of conserved material?
Is it to have material in botanic gardens, or, is it ultimately to
have material to be released?

A reasonable genetic diversity to continue the process over a
period would be necessary for use in a botanic garden, and

diversity is a matter of insurance even under protected conditions. If the object is re-establishment in the wild state, and such restitution into nature has been effected as mentioned by Dr Raven, then the genetic diversity is so much more important. The genetic diversity should be very substantial if there is to be any chance of success.

Here I list a number of guidelines:

1. The main emphasis should be on ecologically suited taxa in any one environment - taxa which you want to reinstate or which you expect to be maintained in your gardens over long periods; not referring to the specimens of highly endangered plants preserved in very small numbers in a glasshouse or garden, where there may be only a handful throughout the world. I do not think such specimens are likely to be preserved over very long periods. In my thinking I am considering taxa which you regard as liable to be maintained over long periods.

2. Any botanic garden should restrict the number of taxa which it seriously wishes to protect. It would not be compatible with the conditions already outlined of space etc. to visualise a large number of taxa conserved in any one institution.

3. Provide genetic diversity within the agreed taxa. This can be done in a garden but better along ecological transects. If you wish to preserve a group of species of any one genus, or even a single species with a widespread distribution, it is far more effectively reserved over an ecological transect. The diversity of ecological conditions of botanic gardens would greatly enhance chances of success. The famous botanic gardens in Bogor has such a transect of a latitudinal nature, with branch stations in Java covering a range of conditions. With a number of gardens in the five Asian countries it may be possible to establish such a co-ordinated scheme. In discussing this with Dr Roy Taylor of the University of British Columbia, I found that he was quite enthusiastic for such an enterprise in South America.

4. Study the breeding system if not known; acquire a good deal of information on autecological aspects of the taxa and the synecology of the community in which it is likely to be established.

5. Provide a baseline data on the genetic diversity of the population to be released. This is highly important. It has not been done in a single instance of a species being

released into nature. Not only is it a matter of great
interest whether under natural conditions a population
will increase or decrease the range of variation, but it
would provide guidance for the conservation problems in
nature reserves, where we have next to no information on
population size and diversity which is required for
survival of organisms. Our only guidance comes from
islands of known ages, where a little information about
lizards in the Mediterranean is known, but there are no
population censuses. We have such censuses for islands of
known age off the coast of Western Australia for mammalian
species known on the mainland.

If a population of plants is released into nature it would
be essential for population geneticists to make a survey
of gene frequencies which is now a relatively easy matter.

6. Finally in such cases one should follow the population
   dynamics after release - both numbers and genetic diversity.

I believe that such an attempt at conservation would strengthen
the chance of continuity in botanic gardens. Where there is a
strong and pervading research interest there are likely to be some
who will fight for continuation. All is insecure. Any national
park secured by acts of parliament is liable to be destroyed with
one signature.

The only way in which we can hope to secure our conservation
efforts of today or tomorrow is by building in interest and by
building up the public awareness and the public conscience - in
fact, by building up an evolutionary ethic and responsibility.

The concept of a time scale of concern has been suggested; it
is the period over which you think an action, an instruction or a
measure should be followed. In the conservation efforts of botanic
gardens the time scale of concern has been too brief. Such as
perhaps their life time, the life time of a tree or of some arrangement made. Unless you can build a time scale into your thinking
and acting in botanic gardens, your efforts are bound to be
ephemeral and limited by the efforts of people. One director may
be keen on a project and the next may throw it out. This has
happened. I believe that the most important thing is for those of
us who are concerned with conservation to have a clearly understood
time scale of concern in our own minds, and change our policies,
thinking and actions to make this time scale a valid and effective
one.

REPORT OF THE INTERNATIONAL ASSOCIATION OF BOTANIC GARDENS PLENARY

SESSION - MOSCOW 1975

P. I. Lapin

Main Botanic Garden

Moscow, Union of Soviet Socialist Republics

The Plenary Session of the International Association of Botanic Gardens (IABG) took place in Moscow at the Main Botanic Garden of the USSR Academy of Sciences on 30th June to 1st July 1975.

The experience of the fifteen years work of the IABG shows that this organisation has contributed essentially to the increase of the theoretical level and practical efficacy of the botanic gardens of the world, and has promoted international co-operation between botanic gardens and arboreta. It has already been recognized that it is impossible in modern times to put forward and solve the great problems of the age particularly the replication of the floristic riches of the Earth, without such co-operation.

The IABG Plenary Session took place under the Chairmanship of its President - Academician N.V. Tsitsin (USSR); the former President of IABG, Dr R. Howard - Director of Arnold Arboretum (USA) also participated.

In accordance with the Constitution of IABG its Plenary Sittings precede International Botanical Congresses and International Congresses on Horticulture. In the past, these meetings took place in parallel with the sessions of these Congresses, but it was found that this impeded the participation of delegates in IABG work. On this occasion the session was held on the three days preceding the opening of the Botanical Congress and thus botanic garden representatives were able to focus their attention on their particular problems and points of interest.

The programme of the Plenary Session was prepared in advance,
after consultation with prominent active members of IABG. It was
decided to discuss the theory and methods of plant introduction;
the most pressing problems of botanic gardens on conservation of
rare and disappearing species of plants; development of proposals
on the improvement of scientific co-operation between botanic
gardens of the world and exchange information in general on the
dissemination of knowledge about plants.

Academician N.V. Tsitsin opened the IABG Plenary Session.
Then, on behalf of Moscow Council of Workers' Deputies, the participants in the Plenary Session were greeted by V.A. Ermolov, the
Deputy of the Moscow Council, Chairman of the Executive Committee
of the Kirov district of Moscow.

He expressed satisfaction that such an important scientific
meeting was being held in the capital of the Soviet State and that
the Main Botanic Garden, USSR Academy of Sciences was the host.
He noted that the Moscow Botanic Garden in its thirty years of
existence had made a profound contribution to biological science,
and had gathered a unique collection of plants, had built very
instructive botanical expositions and had become a centre for the
popularization of new plants valuable for the national economy.

The Main Botanic Garden takes an active part in beautifying
the landscape of Moscow by providing new, useful ornamental plants
and very valuable advice. It was stressed that the Moscow Council
appreciated highly this activity and considered the garden a
remarkable asset to the city.

The plans for further development by the Moscow Council were
in hand, for this work forty-two hectares of new land will be
added to the gardens. In the centre of this area a unique structure
will be built – the climatron – which will be up to thirty metres in
height and covering an area of about one hectare. It will accom-
modate plants of tropical and subtropical countries; individual
plants and whole plant landscapes will be exhibited. A large
twenty-two hectares water reservoir will be built; the main
entrance to the garden and a new underground station will also be
constructed in this new ground.

The IABG General Secretary, Mr D. Henderson, reported on the
activities of the International Association of Botanical Gardens.
The report covering the last six years, as well as the activity of
the Committee was approved.

The scientific programme of the Plenary Session was then
opened with a paper delivered by Academician N.V. Tsitsin entitled
"The Tasks of Botanic Gardens in Conservation of Plant Life".
Vegetation, during many millions of years of evolution has been a

source of nutrition, raw materials and as the means for maintaining
purity of atmosphere and water, as well as being a primary factor
in providing favourable emotional effects on man. Today it has
become even more important. So this report served as a guideline
for the work of the whole Session and affected the resolutions
adopted.

Dr S.M. Walters, Director of the Cambridge University Botanic
Garden, England, presented a paper entitled "The Role of Botanic
Gardens in Conservation of Rare and Disappearing Species of Plants".
In his report Dr Walters concentrated on the problem of the
creation of specially protected plant collections in botanic gardens
paying particular attention to exact documentation of the origin,
environmental conditions, and genetic structure of plants, and the
possibility of using them in hybridization. He suggested that
each botanic garden of the world should specialize in growing a
certain group of endangered species. Such an aim is quite possible
with the help of IABG. He stressed that botanic gardens of the
world are a great force capable of rendering an effective contribution for plant conservation in the next decade.

A.M. Grodzinsky, Director of the Central Republic Botanic
Garden of the UkSSR Academy of Sciences (Kiev) made a report
entitled "Biogeocenology and Plant Introduction". He attached great
importance to the effect of introductions made by man on changing
the vegetation of the biosphere, emphasising the process of
spontaneous introduction of weed species which forced out indigenous
plants including those which are rare and sometimes useful to man.
He gave a few typical examples of spontaneous naturalization.
Asclepias syriaca L. introduced from North America grows well under
the conditions of Kiev and is spreading in the region.
Heracleum sosnowskyi Manden and Crambe cordifolia Stev. introduced
from the Caucasus also grow well, and are actively forcing out
native species.

The author also covered other areas which are of importance to
botanic gardens:

1. Protection and widening of populations of rare and endangered
   species of native floras both wild and cultivated.

2. Elaboration of techniques of cultivation of lasting, resistant plant groups of native and foreign floras on the
   basis of their allelopathic activity and tolerance.

R. Howard, Director of the Arnold Arboretum, USA, spoke on
"Computers in Registration and Analysis in Plant Introduction".
He summarized ten years' experience with some of the USA botanic
gardens in collecting, protecting and studying introduced plants
with the help of computers. There can be no doubt that this

efficient method of registering data on the origin and behaviour of
plants in new environments and with a great diversity of botanical
collections, will be used widely and will promote more effective and
more profound scientific publications in the field of introduction.

S. Bialobok, Director of Arboretum Korniku, Poznania, Poland,
in his report "The Role of Botanic Gardens in the Development of
Biological Science, Economy and Culture of Modern Society", showed
the history of botanic gardens, their achievements in the development of biological sciences, aesthetic and cultural education of
people throughout the world. He illustrated the differing roles
botanic gardens play in scientific research which was dependent
upon their affiliation and purposes, and stressed the great contribution made by the botanic gardens of the world to the development
of biological science, protection and enrichment of vegetation
resources and popularisation of scientific knowledge.

F. N. Rusanov, Director of the Botanic Garden of the Uzbek,
SSR (Tashkent, USSR), in his report "The Principles and Methods of
Studying Living Plants in Botanic Gardens", gave in detail the
method for the study of introduced generic complexes. This method
provides many possibilities for comprehensive comparative study of
introduced plants and selection of the most suitable species for
cultivation. He stressed the basis of this method which can be
used for selection and cultivation in new environments of species
having wide ecological amplitude. If the plants have a narrow
ecological amplitude it is necessary, according to Rusanov, to use
hybridization.

To prove this, Rusanov mentioned the successful work done at
the Tashkent Botanic Garden in hybridization of Hibiscus and Yucca.
Rusanov emphasized the importance of this method in specification
of phylogenic relations and taxonomy of experimental plants.

F. Bencat, Director of the Arboretum in Mlynany, Nitra,
Czechoslovakia, made a report "Organization of International
Botanical Expeditions on the Basis of Mutual Exchange". He reviewed
the history of botanical expeditions and stated that each expedition
should have a special purpose for collecting specific botanical
material, and the best expedition also studied the natural
environment of the species to be introduced. This method provided
the most valuable material for really scientific experiments on
plant introduction. In this connection he emphasized the need to
plan such expeditions properly and suggested that the Session
should elaborate plans for international co-operation in this
sphere. Such projects should include the organization of expeditions on the basis of discussions and agreements among the botanic
gardens of the USSR, USA, Great Britain, Japan, India, Germany and
other interested countries; the expenses to be shared amongst
participants.

R. Seibert, Director of Longwood Gardens, USA, presented the
report "Educational Services of USA Botanic Gardens and Dissemination
of Knowledge among the People". He reported the experience of the
USA botanic gardens in this activity and stressed the necessity of
extending this work in all ways possible. He mentioned the need to
vary the form and content of information provided to fit in with
different categories of visitors (adults, children, amateurs,
students, specialists, etc.).

P. Lapin, Deputy Director, Main Botanic Garden, USSR Academy
of Sciences, Moscow, presented a paper to mark thirty years of
this famous garden's existence, covering all the activities of the
institution.

He said that at the present time the Soviet Union has 115
botanic gardens. Some of them were founded more than two centuries
ago, others are going through the initial stages of their development. On the whole, they have made a tremendous contribution to
the development of botany and have greatly enriched the plant
resources of the USSR. But among these scientific institutions,
there is one which occupies a special place in their work, that is
the Main Botanic Garden of the USSR which was founded on the
14th April 1945.

The decision was taken to construct a unique scientific
institution, a botanic garden of a new type, which was to combine
practical activities in mobilization and exploitation of plant
resources with profound and complex theoretical elaboration of the
problems of plant introduction and acclimatization, using all the
methods available in modern biology.

During the first stages of development of the garden, great
attention was focussed on accumulating botanical collections as a
basis for research work as well as for the creation of botanical
displays. The scientists of other Soviet botanic gardens and those
of foreign countries have helped in gathering seeds and plants for
the collections. The Main Botanic Garden now exchanges seed with
600 botanic gardens and other research and experimental institutions
in the USSR as well as with 750 botanic gardens and arboreta from
80 foreign countries. In this way the garden succeeded in gathering
rich collections and in developing interesting expositions in the
various departments, particularly with regard to flora of the USSR,
dendrology, cultivated plants, floriculture and tropical plants
which between them occupy 360 hectares. The list of botanical
collections in the garden now runs to 20,000 names of plants
originating from all over the world.

The problems of introduction are being studied very extensively
at the Main Botanic Garden with regard to their theoretical and
economic importance using every method of modern biology. Dr Lapin

explained this work has been summarized in a number of fundamental
monographs. Scientists of the garden have published over 4,400
scientific publications on the important problems of plant introduction and acclimatization. The Bulletin of the Main Botanic
Garden is published quarterly in which scientists of the garden
and other botanic gardens of the USSR report the results of their
experimental work. The greater part of this publication makes an
important contribution to the theory of plant introduction and
serves as the basis for the enrichment of the flora and for the
preservation of the plant gene pool.

Dr Lapin said it is difficult to over-estimate the importance
of the Main Botanic Garden in the unification and co-ordination of
all the other botanic gardens of the USSR. For this purpose, the
Council of the botanic gardens of the USSR was organized in 1952.
The Main Botanic Garden with its unique collection of living plants
has become a national treasury of great scientific and cultural
importance, and is actively involved in the protection of rare and
endangered species.

GENERAL CONCLUSIONS

As a result of open discussions a decision was adopted which
took the form of a friendly appeal to all workers in the botanic
gardens of the world. It was an appeal for co-operation and active
participation in the solution of the vital problems discussed at
the sittings of the Plenary Session. This message once again
stated that environmental protection and restoration of its
relative equilibrium and protection of the reproducing potential
of nature at the present time is one of the most acute social,
economic, legislative, scientific and scientific-organizational
problems of our age.

It was emphasized that for the past decade we have achieved
certain success in the conservation and restoration of plant life.
Many species of plants and plant communities are protected within
the boundaries of reservations, national parks, nature reserves,
etc. Certain forms and species of plants are protected by legislation, and international links are maintained and expanded for
the conservation of plant life.

Botanic gardens have started to co-ordinate their work with
the International Union for the Conservation of Nature and Natural
Resources (IUCN), which is an important international organisation
directing the efforts of scientists and specialists towards the
solution of the most pressing problems of environmental protection.
However, these measures are not sufficient and we have much to do
in the near future.

The participants in the Plenary Session recognised that numerous botanic gardens of the world, united under the auspices of the IABG represent a powerful force, which is capable of rendering effective assistance in the conservation of plant life in all continents.

It is evident that the success of preventive activity depends greatly on the perfection of legislation, and on the activity of administrative and social bodies to implement this legislation. It also depends on the level of the awareness of the world's population. The Session recommended that all the botanic gardens of the world, research as well as cultural and educational, should make every effort towards the solution of the above-mentioned problems by the promotion and maintenance of effective legislation by conducting systematic activity on the improvement of the cultural level of the population, and by disseminating ideas of plant life conservation.

It was also considered necessary to unite efforts at national and international level so as to develop a unified method of registration and to the development of criteria and categories, which should be used as guidelines when referring species and plant forms to rare and disappearing plant categories. It was recommended that the IUCN classification of rare and disappearing species of plants should be used. Using these criteria all should participate in the compilation and publication of regional, national and global lists and render every possible promotion to legislative documentation on the protection of the above-mentioned plants.

Botanic gardens must take a leading place in the field of cultivation of rare and disappearing species of plants. This activity should be promoted further and should include:

1. The study in nature and in cultivation of the biology and ecology of rare and disappearing species of plants and the development of methods for their cultivation.

2. Cultivation in botanic gardens of rare and disappearing species of the local flora in combination with the protection of their natural habitat.

3. Compilation and publication of lists and catalogues of rare plants cultivated in botanic gardens and arboreta or growing in natural plant communities located in or near them.

4. Selection of especially valuable and rare species and forms from the local flora, with regard to which botanic gardens or other similar institutions should maintain systematic supervision over their condition and take responsibility for their safekeeping.

The practical implementation of the above-mentioned tasks requires the establishment of the regular exchange of information between botanic gardens of the world and the organization of effective methods for the exchange of seeds and living plants.

Logically, this work should culminate in the artifical re-introduction of disappearing plant communities. The IABG Session recommended that areas with natural vegetation be included within the territory of botanic gardens. In such areas a strict reserva-tion regime should be maintained and systematic research on programmes of conservation of these ecosystems should be under-taken.

The Session recommended botanic gardens should publicise ideas of nature protection and dissemination knowledge of plant life by all possible means: by publishing popular booklets, organization of displays, excursions, production of films and TV programmes, etc.

The successful solution of these extremely important problems requires the extension of international scientific co-operation of botanic gardens and similar institutions by carrying out joint research on co-ordinated programmes, regular holding of inter-national, regional and topical conferences of botanic gardens and arboreta for the discussion of current problems of plant life protection; the organization of joint expeditions by workers of botanic gardens into different geographical areas, etc.

The great value of nature as an initial source of material benefits, as a source of health, happiness, love of life and spiritual wealth of mankind, determines the necessity and priority of the outlined measures in the protection and enrichment of plant life.

The message of IABG emphasizes that the more the riches of nature are used, the richer will be the life of each individual, as the struggle for nature protection is the struggle for the well-being of a human being today, tomorrow and always.

The final act of the Plenary Session was to elect a new IABG Council. Mr T.R.N. Lothian (Adelaide) was elected the President of the IABG. Dr Irwin (New York) and Dr Faegri (Bergen) were elected Vice Presidents. Mr D. Henderson (Edinburgh) was elected General Secretary and the following were elected members of the IABG Council: Professor T.M.C. Taylor (Vancouver), Professor J. Heslop-Harrison (Kew), Professor Dr T. Eckhardt (Berlin), Dr F. Bencat (Czechoslovakia) and Dr P. Raven (Missouri).

The participants of the IABG Plenary Session expressed their appreciation to Academician N.V. Tsitsin for his many years of fruitful activity as the President of the IABG. According to the constitution of the IABG, N.V. Tsitsin became a vice-president of the IABG.

247 participants (delegates and guests) took part in the Plenary Session coming from twenty-five countries: Australia, Argentina, Bulgaria, Great Britain, Hungary, German Democratic Republic, Greece, Israel, Spain, Italy, New Guinea, Norway, Poland, United States of America, Federal German Republic, Finland, France, Czechoslovakia, Sweden, Japan, Cuba, Vietnam, Philippines, Eire and the Netherlands. The Soviet Union was represented by fifteen Constituent Republics.

The documents of the IABG Plenary Session were reported at the twelfth International Botanical Congress and were fully approved by the delegates.

RESOLUTIONS

The IABG Plenary Session calls upon all the workers of botanic gardens of the world actively to participate in nature conservation, in particular in protection of plant life, the most important component of nature.

INTERNATIONAL CO-OPERATION AMONG CONSERVATION-ORIENTATED BOTANICAL GARDENS AND INSTITUTIONS

Edward S. Ayensu

Smithsonian Institute

Washington DC, United States of America

The conservation of nature, particularly of the world's flora, is a problem with complicated solutions at best. During the last few years and certainly in recent months, much effort has been devoted by a number of institutions to inform the world that plants are indeed in danger of their lives and therefore need some concrete forms of protection to ensure their survival.

In 1968, Dr Nicholas Polunin stated that an important prerequisite for the preservation of species in botanical gardens is to recognize which taxa are seriously threatened with extinction. At that time Dr Ronald Melville was beginning work on the angiosperm Red Data Book. We now have not only some treatments in the Red Data Book but a number of lists, such as the Smithsonian Report on Endangered and Threatened Plant Species of the United States and accounts from various countries, such as the United Kingdom, Italy, Peru, the U.S.S.R., Switzerland, and South Africa, and those species on Appendices I and II of the Convention on International Trade in Endangered Species of Wild Fauna and Flora. As has been pointed out in print and verbally by many, fifteen governments have so far ratified the convention. (See Editors' note).

As we know, conservation of plant life was a dominant topic at the Twelfth International Botanical Congress in Leningrad. Earlier, representatives of 247 botanic gardens met in Moscow and they are now pledged to co-operate with the International Union for Conservation of Nature and Natural Resources (IUCN) by taking into cultivation as many rare and endangered taxa as possible to ensure that potential genetic sources will not be completely lost. This process, however, can involve considerable risk as far as

successful propagation of species is concerned; and in this conference there was some debate concerning the necessity to keep these
plants intact in their habitats until rescue was immediately needed.

Dr Peter Raven (this conference) also spoke in some detail
about the ethics and attitudes of plant collecting. I will only
add that the philosophy of leaving plants in their habitat until
the last minute before being rescued is exemplified by the North
Carolina Botanical Garden's rescue team operation at Chapel Hill.
The primary sources of most of the garden's native plants are the
many construction sites across the state of North Carolina. In
1974, the staff, student workers and volunteers rescued, relocated
and conserved thousands of mountain wild flowers and shrubs from a
golf course site at a place called Grandfather Mountain. These
plants would have been bulldozed over if the rescue team had not
been in operation. They have done similar rescue work in areas
projected for drainage, residential development and land-fill.

It seems logical to assume that the botanical institutions of
the world should play a pivotal role in conserving as many as
possible of the plants that are doomed to die because of human
activities, as well as those that seem to be dying a natural death.
This conference will hopefully take a very important step forward
in our attempt to pool our resources together to achieve this goal.
Several suggestions made during the past few days have presented
various methods which, when adhered to, will ultimately lead to
some concrete improvement in the conservation of living plants. I
only hope that we can agree on a set of methods to be followed by
all institutions.

I submit that it will be most unfortunate if only a few
botanical gardens are saddled with the job of rescuing plants that
can no longer be preserved in their natural habitats for one reason
or another. It is my firm conviction that success will come our
way if all botanical institutions, be they large or small, rich or
poor, in temperate regions or in the tropics, can share materials
and information in an organized manner. This positive orientation
towards conservation has been taken by a number of botanical gardens.
Mr K. Woolliams (this conference) has said that the Waimea Arboretum
on Uahu, Hawaii, has as its primary role the cultivation and distribution of endangered plants expressly to form a living gene pool
for use in breeding programmes and other purposes. Plants from
Hawaii, the Canary Islands and Fiji are currently being established
at Waimea. Included in the programme to conserve the endemic
Hawaiian palms in the genus Pritchardia, of which ten species are
listed as endangered in the Smithsonian Report.

The US National Arboretum has also played an important role in
ensuring the survival of the Lost Franklinia (Franklinia alatamaha)
through cultivation and distribution of seeds and plants to botanical
gardens and other interested parties. The plants became extinct in

the wild shortly after the early 1800's. No doubt some of the
major botanical gardens in Europe also have engaged in similar
activity. Apart from botanical institutions, many other scientists
who do not work in botanical gardens will have to become involved.
We need a truly massive front to have a reasonable chance to solve
the problems before us.

I have decided to focus on four major steps which I hope will
help our efforts to achieve the aims inherent in the theme of this
Conference.

The first step to be taken is for all botanical gardens to take
an inventory of their holdings of cultivated endangered species and
to circulate the results internationally. The American Horticultural
Society's Plant Sciences Data Centre has already made significant
progress towards the development of an internationally recognized
information centre, the object of which is to establish a centralized
computerized inventory of cultivated plants within protected collections such as those in botanical gardens, arboreta, and parks of
North America. Approximately twenty-six botanical gardens have had
their plant accession records computerized and processed into the
Data Centre's information bank. This includes the records from
Fairchild Tropical Garden (Florida), Arnold Arboretum (Massachusetts)
Denver Botanic Gardens (Colorado), Longwood Gardens (Delaware),
Morton Arboretum (Illinois), United States National Arboretum
(Washington DC), Los Angeles State and County Arboreta and Botanic
Gardens (California), and the Royal Botanic Garden (Ontario, Canada).

Twenty-five other botanical institutions are using special data
cards developed by the Plant Sciences Data Centre to record their
accessions and are actively involved in the Centre's programme.
The Centre now has over 135,000 records on cultivated plants representing about 50,000 taxa in 3,500 genera reported within protected
collections. In addition, the Centre also services inquiries from
all over the world. The Centre's record management programme includes computerization of the United States Department of Agriculture
(USDA) Fruit Germplasm Resources, USDA Plant Introduction Records,
plant nomenclature cross-references to botanical synonyms in USDA
publications, and International Plant Registration data.

The Centre's System has now received the computer-taped listings of all taxa in the Smithsonian Institution's Report on Endangered and Threatened Plant Species of the Unites States, which
Secretary S. Dillon Ripley presented to the Congress of the United
States at the end of last year; and which on 1st July this year the
Department of the Interior placed on notice of review for public
comment. I am sure you can imagine the potential of the data that
can be obtained from this activity, which will result in our being
able to know just how many of the plants in the Smithsonian Report
are currently being cultivated at participating gardens in the

United States and Canada. These data will be shared with other
institutions and also submitted for inclusion in the International
Union for Conservation of Nature and Natural Resources' Red Data
Sheets under the standard heading "Conservation Measures Taken".

The second specific step that we must take involves research
on the plants about whose biological syndromes and natural habitat
regimes we seem to have not enough information. The lectures presented by Professors Esser, Heslop-Harrison and Cook attest to our
lack of knowledge in these areas. The removal of plants from their
natural habitats into cultivation in order to perpetuate and
efficiently reproduce them could result in a futile exercise, despite regular intensive horticultural care in botanical gardens,
unless research has been done, both in the field and in our institutions, on their reproductive systems.

Work along these lines is also being undertaken at the North
Carolina Botanical Garden, where a major effort is made to learn
more of the cultural requirements and propagation techniques for
rare plants such as the Venus's fly-trap (Dionaea muscipula) and
pitcher plants (Sarracenia). Learning to grow these plants from
seed is helping them to demonstrate to the public that certain
native plants can be propagated relatively easily from seed, and
that entire plants should never, in their opinion, be dug from their
natural habitat unless endangered by man's encroachment on their
environment. Another example of plant reproduction research is the
investigations into the methods of rooting native Hawaiian plants
under controlled conditions at the Pacific Tropical Botanical Garden
at Kauai. In addition, Dr Robert Farmer of the Tennessee Valley
Authority (Norris, Tennessee) is currently working on the propagation of some rare native plants of Tennessee.

Botanical gardens and other research institutions should become
more concerned with the active monitoring of population levels of
these species in the field, in addition to investigating their soil
preference, specific insect pollinators, and reproductive systems.
There are far too many endangered species to leave this task to a
few taxonomists and ecologists not employed by botanical gardens.
Knowledge gained by botanical garden staff about plant population
levels (whether they are increasing, decreasing or stable) and their
reproductive potential in cultivation can be of significant use for
future decision making. For example, the US Department of the
Interior, which has the management authority in the United States,
considers data on population levels essential for their decision as
to the official listing of American plants as endangered or threatened, and thus requiring legal protection under the Endangered
Species Act of 1973. The Department of the Interior is also charged
with the responsibility of determining which American plants (in
addition to those already listed) should be subject to regulation
of their export overseas under provisions on the Convention of
International Trade in Endangered Species of Wild Fauna and Flora.

A knowledge of the true population levels of candidate plants will help them to administer their permit granting wisely. Indeed, high officials in the Department of the Interior have said that their main goal is to take species off the list, as their population levels improve. I think it is our responsibility to find out more about current population levels of the endangered and threatened species in the wild. It would seem logical that the agencies in countries that have designated forest reserves and other natural preserves could take an active part in assessing the population levels of the endangered plant species on their lands.

Most of you are already familiar with the case of the Georgia Plume (<u>Elliottia racemosa</u>, <u>Ericaceae</u>), a rare shrub growing up to fifteen feet tall, originally of the Savannah River Valley in Georgia, USA. It has been difficult to reproduce in cultivation since it is self-sterile and produces very few viable seeds. Furthermore, vegetative reproduction of this plant also has proved difficult, although in nature it forms clumps from its rootstocks. Cross-fertilization from one clump to another growing close by must have been accomplished occasionally before the advent of human civilization. So many habitats containing this plant have been wiped out by lumbering and agriculture so that, as early as 1933, propagation of seeds in the field had seemed to cease. Fortunately, the Georgia Heritage Trust, an agency of the Georgia Department of Natural Resources, has now acquired a 750 acre Pleistocene sand ridge on which grows the largest known surviving population of <u>Elliottia racemosa</u>. This Trust acquired the land for an important reason, ie. to preserve the Georgia Plume in its habitat, where its reproduction can be studied <u>in situ</u>, and then it can be brought successfully into botanical gardens for the enjoyment of many.

Another case in point is the endangered, world-famous <u>Lodoicea maldivica</u> (coco-de-mer; double coconut), occurring on three reserves on Praslin Island in the Seychelles, which I recently visited. Unless its reproductive habits and life history are critically studied in the field in order to be able to suggest methods of cultivating it, I am inclined to conclude that it will be no use to try to breed it in botanical gardens. The double coconut has water-stress problems and an excessively low rate of fruit production. Although I am aware of the many predictions that the double coconut will die out in its natural habitat of its own accord even if completely undisturbed by man, I am also confident that current and future scientific knowledge, combined with our willingness to save this plant, can negate the negative predictions and bring this curious plant to the view of many people.

These are but two examples from the many thousands of plant species that are currently known to be declining in numbers for one reason or another. The point we have to bear in mind is to ensure that we establish a systematic method for sharing our research information with the 600-odd botanical gardens in the world today.

The third specific act that botanical gardens the world over can engage in is the exchange and sharing of materials in the form of seeds, seedlings and cuttings. We are all familiar with examples of plants that often do poorly in their native habitat but seem to do rather well when cultivated in other geographical regions across national boundaries. The major European and North American botanical gardens have had rather long historical connections with many of the botanical gardens in the less-developed world. I would like to submit that the major botanical gardens should use every influence at their command to get such an endangered species exchange programme going between them and particularly, the botanical gardens in the tropical regions of the world. For example, a letter from a major botanical garden to a commonwealth country in the tropics could serve two important purposes. Firstly, the latter director will consider a letter from the director of a major garden requesting some involvement in this programme as an important note of recognition and the kind of ethos needed to get local governmental backing. Secondly, the exchange of the plant materials between the developed and under-developed botanical gardens and institutions will help to increase the survival chances of many plant species that are endangered.

A number of the contributors to these proceedings have expressed, in various ways, the need for all of us to get involved in the tropics. I think I will be failing in my duty if I do not make a comment or two in support of the suggestions that have already been made. We must first recognize that the genetic resources in tropical plants should not be viewed as intrinsically the sole property of the less-developed countries. On the contrary, it is the responsibility of all of us to ensure that we protect and perpetuate the tropical plants because all our lives are tied up in such resources.

Most of the tropical countries are currently in serious financial straits. Hence it is not difficult for us to see why very little funds are being directed into the establishment and/or maintenance of botanical gardens and national parks. In fact, because of the desperate attempts by the developing countries to meet their food bills, the very areas that we wish to see preserved and maintained are the same prime lands that are being destroyed in order to grow some food to support their populations.

Some of the less-developed countries get very upset if they are told to preserve their environment in the face of their financial crises. One African diplomat remarked to me at the United Nations in New York City some weeks ago that most developing countries find it difficult to understand why the affluent countries are religiously preaching that poor countries should develop large forest reserves.

After all, the developed countries have exploited their own environments to the full and are now rich. Why cannot the poor nations do
the same and get rich before they can talk about conservation?
Manufactured goods such as farm machinery, are not getting any
cheaper. Therefore, the poor nations should continue to cut their
timber for sale in order to meet their balance of payment problems.
This attitude generally describes the feelings of most developing
countries.

It is therefore our collective responsibility to encourage
developing countries to seek international funding, plus local
matching funds, for the establishment of botanical gardens and
parks. The first phase of such an effort should be directed to the
establishment of national parks since the initial financial outlay
in their upkeep may not be as prohibitive as the establishment of
a full-scale botanical garden. It would also seem that a national
park in the tropics may be relatively easy to manage with a small
policing force.

Furthermore, a new kind of partnership should be developed
between the gardens of the developed countries and those of the
less-developed countries. I would like to caution, however, that
the new kind of partnership should not carry with it the trappings
of the old colonial relationships. We shall defeat the aims of
this conference if we do not make a concerted effort to build true
co-operation among the gardens. If this is done, we shall see the
upgrading of the status of botanical gardens in the less-developed
countries. We shall, in fact, help the developing countries to
attract some of their highly trained scientists to get involved in
the administration of their tropical gardens.

To digress further, I would like to mention that during a recent visit to the People's Republic of China, I was gratified to
learn that our colleagues in Chinese botanical gardens and forestry
departments are very much interested in the conservation of living
plants. The director of Kwangtung Botanical Garden in Kwangchow
(Canton), Professor F.H. Chen, informed me that in their plant introduction programme they have been concentrating not only on medicinal
plants collected from many parts of China and other parts of the
world, but plant species that seem to be in danger of extinction.
The Kwangtung Botanical Garden is also taking an active part in the
China Conservation Committee. The Chinese conservation programme
was inaugurated last year between the end of August and the first
week of September in Harbin (a city in north-eastern China) in
Heilungkiang Province near Siberia. As a result of the conference,
a number of research institutes and conservation stations have been
established. In Yunnan Province, for example, four such stations
have been established to study the plants in a 6,000 hectare reserve.
In Kwangtung Province there is another station about eighty kilometers outside Canton. The conservation station is under the dual

jurisdiction of the Kwangtung Institute of Botany and the Administration of Kwangtung Province. The discovery of a new species of
silver pine in Kwangsi Province and some new plant genera resulting
from extensive plant explorations, as well as the discovery of new
conifer groves of Ginkgo and Metasequoia, has given a new impetus
to their conservation programme. I am pleased to inform you that
our colleagues in China are willing to exchange material and information with their counterparts abroad.

A fourth specific activity which botanical gardens can engage
in is to devise a method by which information on their research and
holdings of endangered plants can be communicated to people doing
vital work on the subject in institutions at local, state and
governmental levels around the world, but who simply do not have
easy access to literature circulated among botanical gardens themselves. There are perhaps already too many new journals, but a
newsletter on this subject coming from an appropriate botanical
body appears to be necessary.

A combination of forceful yet rational legislation to protect
endangered plant species, coupled with balanced ideas as to when
to rescue them from their habitats, vigorous involvement in basic
research on their biology, and the preservation of them in cultivation in botanical gardens as propagation sources, hopefully will
result in their being eventually preserved for all of mankind.

The cycads may be cited as an instance where such interplay
between wise legislation and conservation in botanical gardens could
be the only way to save their wild populations from extinction. The
crux of the cycad situation has been explained by Dr Gilbert Daniels
in the following way: because cycads are relatively rare, slow to
reproduce and take many years to reach maturity, it is essential
that legislation for their protection be soundly based. It is
Dr Daniels' contention that an overall embargo on the sale, transport, importation or exportation of these plants may not be
altogether reasonable. He believes that legislation to protect the
plants should allow for their seed to be collected, propagated and
re-established in the wild. The wild populations in Central America,
Australia and Fiji are composed almost entirely of mature stands,
since the clumps of seedlings found beneath female plants rapidly
die off due to overcrowding. I have also observed a similar situation occurring with the wild populations of Encepharlartos barteri
in Ghana. The seeds should, therefore, be collected, grown in
botanical gardens and the young plants replanted in the wild.

In conclusion I would like to suggest that the authorities of
botanical gardens should not consider their own holdings as the
only plant material that should require their attention and care.
The world is a magnificent botanical garden and all the botanical
institutions are equally responsible for its proper maintenance.

Achievement of this goal will require the co-operation of all the botanical gardens of the world. It is my hope that this conference will take practical steps to involve each of the botanical gardens in our attempt to conserve the world's flora.

Editors' Note. The following countries have now ratified or acceded to the Convention. In order of ratification they are:

> United States of America
> Nigeria
> Switzerland
> Tunisia
> Sweden
> Cyprus
> United Arab Emirates
> Chile
> Ecuador
> Uruguay
> Canada
> Mauritius
> Nepal
> Peru
> Costa Rica
> South Africa
> Brazil
> Malagasy Republic
> Niger
> German Democratic Republic
> Morocco
> Ghana
> Papua New Guinea

REFERENCES

AYENSU, E.S. (1975). Endangered and threatened orchids of the United States. American Orchid Society Bulletin 44(5): 384-394.

BENSON, L. (1975). Cacti-bizarre, beautiful, but in danger. National Parks and Conservation Magazine 49(7): 17-21.

BROWN, R. (1973). Plants in the computer. American Horticulturist 52(4): 36-42.

[Editor]. (1975). How to save a wild flower. National Parks and Conservation Magazine 49(4): 10-14.

FOSBERG, F.R. (1971). Endangered island plants. Bulletin of the Pacific Tropical Botanical Garden 1(3): 1-7.

FOSBERG, F.R. (1972). Our native plants - a look to the future. National Parks & Conservation Magazine 46(11): 17-21.

GWYNNE, P. and GOSNELL, M. (1975). Fading flowers.   Newsweek 86(2): 72.

HESLOP-HARRISON, J. (1973). The plant kingdom: an exhaustible resource?   Transactions of the Botanical Society of Edinburgh 42: 1-15.

HUNT, D.R., (ed.) (1974). Succulents in peril.   Supplement to International Organization for Succulent Plant Study Bulletin 3(2): 1-32.

HUXLEY, A. (1974). The ethics of plant collecting.   Journal of the Royal Horticultural Society 99: 242-249.

ILTIS, H. (1969). The conservation of nature: whose problem? Bioscience 23: 248-253.

JENKINS, D.W. (1975). At last - a brighter outlook for endangered plants.   National Parks & Conservation Magazine 49(1): 13-17.

JENKINS, D.W. and AYENSU, E.S. (1975). One-tenth of our plant species may not survive.   Smithsonian 5(10): 92-96.

LAYNE, E.N. (1973). Who will save the Cacti? Audubon 75(4): 4.

LITTLE, E.L. Jr. (1975). Our rare and endangered trees.   American Forests 81(7): 16-21, 55-57.

LYONS, G. (1972). Conservation: a waste of time?   Cactus & Succulent Journal(U.S.A.) 44(4): 173-177.

MELVILLE, R. (1970). Red Data Book, Volume 5 - Angiospermae. International Union for Conservation of Nature and Natural Resources.   Morges, Switzerland.

MELVILLE, R. (1970). Plant conservation in relation to horticulture. Journal of the Royal Horticultural Society 95(11): 473-480.

MOORE, J.K. and BELL, C.R. (1974). The North Carolina Botanical Garden - a natural garden of native plants.   American Horticulturist 53(5): 23-29.

MORTON, J.K. (1972). The role of botanic gardens in conservation of species and genetic material.   In RICE, P.F. (ed.) Proceedings of the Symposium - a national botanical garden system for Canada.   Royal Botanical Gardens, Hamilton, Technical Bulletin 6: 1-9.

PERRING, F. (1975). Plant conservation – without botanists? New Scientist 67(959): 194-195.

PICKOFF, L.J. (1975). Our role in conservation. Cactus & Succulent Journal (U.S.A.) 47(1): 20-22.

POLUNIN, N. (1968). Conservational significance of botanical gardens. Biological Conservation 1(1): 104-105.

PRESTON, D.J. (1975). Endangered plants. American Forests 81(4): 8-11, 46-47.

ROWLEY, G. (1973). Save the succulents ! - practical step to aid conservation. Cactus & Succulent Journal (U.S.A.) 45(1): 8-11.

SMITHSONIAN INSTITUTION. (1975). Report on Endangered and Threatened Plant Species of the United States. House Document No. 94-51. U.S. Government Printing Office, Washington, D.C.

WALTERS, S.M. (1973). The role of botanic gardens in conservation. Journal of the Royal Horticultural Society 98(7): 311-315.

WEINBERG, J.H. (1975). Botanocrats and the fading flora. Science News 108(6): 92-95.

CONSERVATION: RECENT DEVELOPMENTS IN INTERNATIONAL CO-OPERATION

AND LEGISLATION

G. Ll. Lucaş

Royal Botanic Gardens, Kew

England

There are two relatively new and far-reaching developments of merit in plant conservation on which I wish to elaborate in this paper.

First the setting up of the Threatened Plants Committee (TPC) in mid-1974 by the Survival Service Commission of the International Union for Conservation of Nature and Natural Resources. This was the logical outcome of the work started by Dr R Melville, based at Kew, on the "Red Data Book for Threatened Plant Species". It became obvious from the calculations made on the basis of this preliminary information that about ten per cent of the world's flowering plants were threatened in some way - this means 20-25,000 species. The logical step, therefore, was to attempt to enlist the help of every conservation-minded botanist in the world, firstly to find out which were the threatened species and secondly to do something about their conservation, based on the information received. Thus the Survival Service Commission set up a small working party that was asked to report on how the information gathering and follow up could be best done in the quickest possible way. The Report suggested the TPC structure and Professor J Heslop-Harrison, Director of the Royal Botanic Gardens, Kew, was asked to become its first Chairman and to sit on the Survival Service Commission.

The basic structure of the Committee is in three parts and is best summarized by the Chairman's own paper (Heslop-Harrison 1974).

1. Regional. Groups throughout the world, botanists, horticulturists, ecologists and others with special interest in
   and knowledge of their regional floras, are currently being
   asked to become associated with the work of the Committee,

271

and to form regional subcommittees which will take special
responsibility for identifying threats to floras, plant
groups and individual species in their own areas.   Part of
their task will be to document adverse changes affecting
floras and species, and to prepare recommendations for
minimizing their effect through governmental or other action.
They may be expected also to be involved in protection and
rescue operations.

2. Specialist groups.   To complement the work of the regional
   TPC organization, panels are to be set up to handle the
   problems of special groups - for example the palms, orchids
   and cycads.   In part these groups will have a taxonomic
   basis, and specialists will be invited accordingly;   but it
   is anticipated that panels will be formed to deal with more
   comprehensive categories, based upon ecology or life form.
   Succulent plants could ultimately form such a category.

3. Institutional.   The third component in the TPC will be
   made up of botanic gardens, university departments, research
   institutes and other bodies which have the facilities and
   expertise to maintain plants in cultivation or in seed banks.
   Already the Food and Agriculture Organization of the UN is
   organizing a world-wide network of plant genetic resource
   centres for the husbanding of cultivated species and some of
   their wild relatives.   The aim now is to set up a complementary network to handle natural-source plant material not
   already covered by the activities of the FAO system.

It is under the third heading that a major role for the forward-
looking botanic gardens of the world exists.   However, I shall say
a little about the other two headings first.

THE REGIONAL STRUCTURE

We gather information at a country or region level from the
botanist or botanists most concerned with the conservation of their
flora.   We record the species on cards and build up country lists.
We then have the rather difficult task of assessing on a world basis
the degree of threat to the listed species.   There are many easily
categorised species that are of limited distribution and threatened
by developments which will cause their extinction or severely
diminish their range in the wild.   On the other hand there are more
widespread species which while badly threatened in one country are
far less threatened in another.

Here we have political as well as conservation criteria to
take into account before a world assessment can be made.   However,
without that assessment how can it be said "yes, this is an endangered species and we must take extra care to ensure its survival

and successful propagation"?

We have in North America under the Chairmanship of Dr E. Ayensu of the Smithsonian Institution a Committee that has already (through the Smithsonian) produced for the American Government the first list of endangered and threatened species of the USA. In Europe a preliminary list has been produced by the TPC Secretariat in preparation for a list to be submitted next year to the Council of Europe, the Chairman of the European Committee being Dr S.M. Walters of Cambridge. In nearly all parts of the world we are building up our Committees, not to sit and talk but to gather information by correspondence and by any means available to help present the world-wide picture. We can only succeed with your help and support, so please HELP! Without information none of us can make rational decisions about conservation, but with lists of threatened species we can select those plants most suited to conservation in the wild or in our botanic gardens, so that they can be propagated and held to ensure their future survival. Although 25,000 species is a large number, spread among the botanic gardens of the world it becomes a manageable problem which we can all do something about.

THE SPECIALIST GROUPS

The Specialist Groups, for instance the Succulent, Cycad, Tree-fern and Palm Groups, have a particularly important role in that they bring together experts in these groups to advise on a world basis on the species threatened, the threats, and to suggest remedial action. We have in these Specialist Groups an overlying network to ensure that the major plant taxa which enter trade from the wild will be well documented in the shortest possible time. Compared with the animal "Red Data Books" our information gathering is over fifteen years behind - and yet the pace of destruction is ever increasing.

The TPC has a large number of dedicated botanists from all over the world co-operating with the Secretariat at Kew, but many areas remain uncovered so all help is welcomed by Mr Synge who is employed full-time by the IUCN/WWF to help in this co-operative venture, and by myself acting as Secretary. If there is a group or region which is your special interest and there are conservation problems for the plant species, we want to know. It is only in this way that scientifically based decisions can be made on how best to save these plants; whether by the setting up of local reserves or transferring the plants to protected areas, or going that one step further, taking them into botanic gardens to propagate and maybe reintroduce to the wild. For many species none of these options will be possible and we will witness their extinction. I fear many forest trees will go this way. There is

another basic reason for needing to know which species as well as
which ecosystems are endangered - again almost a political point.
If a country has, say, thirty-five endemic species threatened in
a small area it provides a far stronger case for the local
conservationists if they can say to their Government, whether
local or national, "These species are unique in the world and it
is our responsibility to ensure they are not lost for all time".
Many politicians respond well to this idea of having the ultimate
responsibility for their endemic species.

THE INSTITUTIONAL BODIES

We come then to the third group, those institutions etc. that
have facilities to maintain plants in cultivation or in seed banks.
The TPC hope that this is the role with which the Kew Conference
delegates will become involved, in fact without such help a lot of
our information gathering and future recommendations will become
sterile academic exercises while the species die out. So for the
second time, HELP! - this time to grow and store the species
most suited to your institution's resources. The aim of the TPC
is firstly to provide information - I hope all delegates and
others will use us in this way, with TPC acting as a servicing
agency for all information on plant conservation at species level.
We are sure that the best and cheapest place to conserve species
is in their original habitat which should be protected.

This will not be possible in some cases and so here is a major
reason for botanic gardens to be in close touch, with us and with
each other. Then we can say which species need the protection and
aid of gardens. Secondly, there are many species, very rare in
the wild, which therefore become collectors items. Most of these
can be propagated in large numbers, particularly by the horticultural trade, so the pressures on the limited wild population
can be eased. The importance of documentation has already been
established, and for conservation purposes it is vital so that
separate populations of the same species can be maintained to
ensure wide genetic variability in the stocks held.

This brings me to my final point on the structure of TPC -
we are gathering information about which species are threatened
and I hope we will be able to play a major role in co-ordinating
information on what you are able to grow from these lists, so
pinpointing areas where as botanic gardens we need to concentrate
our energies to ensure the survival of the maximum number of species
for the future. We will not be able to "save" them all but at
least we should know which to concentrate our resources on first.

THE CONVENTION ON INTERNATIONAL TRADE IN ENDANGERED SPECIES

The other major development in plant conservation which I want
to say something about occurred in Washington in early 1973, when
the Convention on International Trade in Endangered Species of Wild
Fauna and Flora was signed. The Conference which prepared the
Convention was attended by representatives from over eighty
countries; some twenty-three of which have so far ratified or
acceded to the Convention which came into force on 25th August,
1975.

The Convention has two main aims: (a) to monitor those species
or groups of species in which trade occurs mainly from the wild-
taken stock, and (b) control of such trade where it is a threat to
the survival of those species in the wild. Hence there are two
Appendices with appropriate degrees of control. The text with all
relevant notes and appendices of plants covered are presented as
an appendix to this paper. In simple terms:

Appendix I includes all species threatened with extinction
which are, or may be, affected by trade.

Appendix II includes (a) all species which, although not
necessarily threatened at present, may become so unless trade is
strictly regulated and (b) species whose trade must be controlled
if trade in the species (a) above is to be regulated effectively.

A third appendix is provided to cover any signatory state's
own particular conservation problems.

Appendix III includes all species designated by any Party to
the Convention because they are protected in the Party's own
territory and the co-operation of other Parties is essential for
the trade to be controlled.

This remarkable Convention indicates that Governments are
willing to aid one another in ensuring that their natural animal
and plant wealth is not squandered for short-term gain and that
conservation is important to them. We must ensure that they have
the knowledge and support to make the Convention work in reality.
The Appendix lists do not represent anything like the real picture
of plant groups which should be monitored and if necessary controlled. Helping the TPC will aid the production of a more
realistic list over the next few years.

The horticultural trade should be encouraged by the Convention
to propagate large quantities from their original wild stocks,
whereas the unscrupulous dealer that continuously takes from the
wild to maintain his trade will find it increasingly difficult to

get export licences for the large plundering collections that he
once used to make. In other words, the legitimate trade should
prosper from these controls. For the bona fide botanic gardens
and research establishments there is an exemption provided under
Article VII.

In order to remove some of the elitism from the list by high-
lighting certain species as really rare and therefore "collectors
items" it will be seen that in some cases genera or families have
been placed on the appendix. This was also done to ensure the
Customs Officer at the port of entry could recognise the group in
the form in which it enters trade and could ensure the appropriate
documentation was available so that illegal transfer of material
was stopped.

From the setting up of the TPC and of the Convention, these
two separate and now intimately linked features, there is now the
opportunity on the one hand to see which species are threatened on
a world scale, why they are threatened and which ones are in trade;
and on the other hand for the botanic gardens of the world to
collaborate in growing the most important of them. I hope that
the TPC lists of threatened species and lists of species held in
cultivation can be brought together, so that we will then be able
to co-ordinate an action programme to fulfil what I am sure is our
most deep-felt wish, to ensure the widest range of plants have a
future.

REFERENCES

[Anonymous.] (1975). A preliminary draft for a list of threatened
    and endemic plants for the countries of Europe. IUCN,
    Threatened Plants Committee Secretariat, Royal Botanic Gardens,
    Kew.

FRANKEL, O.H. & BENNETT, E. (eds.) (1970). Genetic resources
    in plants: their exploration and conservation. IBP Handbook
    No 11, Blackwells Scientific Publications, Oxford.

GOMEZ-POMPA, A., VAZQUEZ-YANES, C. & GUEVARA, S. (1972). The
    tropical rain forest: a non-renewable resource. Science
    177: 762-765.

HESLOP-HARRISON, J. (1974). Postscript: The Threatened Plants
    Committee. In HUNT, D.R. (ed.) Succulents in Peril,
    Supplement to IOS Bulletin 3(3): 30-32.

HESLOP-HARRISON, J. (1974). Genetic resource conservation: the
    end and the means. Journal of Royal Society of Arts, 122:
    157-167.

LUCAS, G.Ll. (1974).   The Convention on Trade in Endangered
    Species.   In HUNT, D.R. (ed.) Succulents in Peril, Supplement
    to IOS Bulletin 3(3): 13-15.

MELVILLE, R. (1970).   Red Data Book, Volume 5 - Angiospermae.
    IUCN, Morges, Switzerland.

CONVENTION ON INTERNATIONAL TRADE IN ENDANGERED SPECIES OF WILD

FAUNA AND FLORA

The Contracting States recognising that wild fauna and flora in their many beautiful and varied forms are an irreplaceable part of the natural systems of the earth which must be protected for this and the generations to come; conscious of the ever-growing value of wild fauna and flora from aesthetic, scientific, cultural, recreational and economic points of view; recognizing that peoples and States are and should be the best protectors of their own wild fauna and flora; recognizing, in addition, that inter-national co-operation is essential for the protection of certain species of wild fauna and flora against over-exploitation through international trade; convinced of the urgency of taking appropriate measures to this end, have agreed as follows:

ARTICLE I

Definitions

For the purpose of the present Convention, unless the context otherwise requires:

(a) "Species" means any species, subspecies, or geographically separate population thereof;

(b) "Specimen" means:

(i) any animal or plant, whether alive or dead

(ii) in the case of an animal: for species included in Appendices I and II, any readily recognizable part or derivative thereof; and for species included in Appendix III, any readily recognizable part or derivative thereof specified in Appendix III in relation to the species; and

(iii) in the case of a plant: for species included in Appendix I, any readily recognizable part or derivative thereof; and for species included in Appendices II and III, any readily recognizable part or derivative thereof specified in Appendices II and III in relation to the species;

(c) "Trade" means export, re-export, import and introduction from the sea;

(d) "Re-export" means export of any specimen that has previously been imported;

(e) "Introduction from the sea" means transportation into a State of specimens of any species which were taken in the marine environment not under the jurisdiction of any State;

(f) "Scientific Authority" means a national scientific authority designated in accordance with Article IX;

(g) "Management Authority" means a national management authority designated in accordance with Article IX;

(h) "Party" means a State for which the present Convention has entered into force.

ARTICLE II

Fundamental Principles

1.   Appendix I shall include all species threatened with extinction which are or may be affected by trade. Trade in specimens of these species must be subject to particularly strict regulation in order not to endanger further their survival and must only be authorized in exceptional circumstances.

2.   Appendix II shall include:

(a)   all species which although not necessarily now threatened with extinction may become so unless trade in specimens of such species is subject to strict regulation in order to avoid utilization incompatible with their survival; and

(b)   other species which must be subject to regulation in order that trade in specimens of certain species referred to in sub-paragraph (a) of this paragraph may be brought under effective control.

3.   Appendix III shall include all species which any Party
      identified as being subject to regulation within its juris-
      diction for the purpose of preventing or restricting
      exploitation, and as needing the co-operation of other parties
      in the control of trade.

4.   The Parties shall not allow trade in specimens of species
      included in Appendices I, II and III except in accordance
      with the provisions of the present Convention.

ARTICLE III

Regulation of Trade in Specimens of Species included in Appendix I

1.   All trade in specimens of species included in Appendix I shall
      be in accordance with the provisions of this Article.

2.   The export of any specimen of a species included in Appendix I
      shall require the prior grant and presentation of an export
      permit. An export permit shall only be granted when the
      following conditions have been met:

      (a)   a Scientific Authority of the State of export has advised
             that such export will not be detrimental to the survival
             of that species;

      (b)   a Management Authority of the State of export is
             satisfied that the specimen was not obtained in con-
             travention of the laws of that State for the protection
             of fauna and flora;

      (c)   a Management Authority of the State of export is
             satisfied that any living specimen will be so prepared
             and shipped as to minimize the risk of injury, damage
             to health or cruel treatment; and

      (d)   a Management Authority of the State of export is
             satisfied that an import permit has been granted for
             the specimen.

3.   The import of any specimen of a species included in Appendix I
      shall require the prior grant and presentation of an import
      permit and either an export permit or a re-export certificate.
      An import permit shall only be granted when the following
      conditions have been met:

      (a)   a Scientific Authority of the State of import has advised
             that an import will be for purposes which are not
             detrimental to the survival of the species involved;

(b)   a Scientific Authority of the State of import is satis-
      fied that the proposed recipient of a living specimen
      is suitably equipped to house and care for it; and

(c)   a Management Authority of the State of import is
      satisfied that the specimen is not to be used for
      primarily commercial purposes.

4.   The re-export of any specimen of a species included in
     Appendix I shall require the prior grant and presentation of
     a re-export certificate. A re-export certificate shall only
     be granted when the following conditions have been met:

(a)   a Management Authority of the State of re-export is
      satisfied that the specimen was imported into that
      State in accordance with the provisions of the present
      Convention;

(b)   a Management Authority of the State of re-export is
      satisfied that any living specimen will be so prepared
      and shipped as to minimize the risk of injury, damage
      to health or cruel treatment; and

(c)   a Management Authority of the state of re-export is
      satisfied that an import permit has been granted for
      any living specimen.

5.   The introduction from the sea of any specimen of a species
     included in Appendix I shall require the prior grant of a
     certificate from a Management Authority of the State of intro-
     duction. A certificate shall only be granted when the follow-
     ing conditions have been met:

(a)   a Scientific Authority of the State of introduction
      advises that the introduction will not be detrimental
      to the survival of the species involved;

(b)   a Management Authority of the State of introduction is
      satisfied that the proposed recipient of a living
      specimen is suitably equipped to house and care for it;
      and

(c)   a Management Authority of the state of introduction is
      satisfied that the specimen is not to be used for
      primarily commercial purposes.

ARTICLE IV

Regulation of Trade in Specimens of Species included in Appendix II

1.   All trade in specimens of species included in Appendix II shall
      be in accordance with the provisions of this Article.

2.   The export of any specimen of a species included in
      Appendix II shall require the prior grant and presentation of
      an export permit. An export permit shall only be granted when
      the following conditions have been met:

      (a)   a Scientific Authority of the State of export has
             advised that such export will not be detrimental to the
             survival of that species;

      (b)   a Management Authority of the State of export is
             satisfied that the specimen was not obtained in contra-
             vention of the laws of that State for the protection of
             fauna and flora; and

      (c)   a Management Authority of the State of export is
             satisfied that any living specimen will be so prepared
             and shipped as to minimize the risk of injury, damage
             to health or cruel treatment.

3.   A Scientific Authority in each Party shall monitor both the
      export permits granted by that State for specimens of species
      included in Appendix II and the actual exports of such speci-
      mens. Whenever a Scientific Authority determines that the
      export of specimens of any such species should be limited in
      order to maintain that species throughout its range at a
      level consistent with its role in the ecosystems in which it
      occurs and well above the level at which that species might
      become eligible for inclusion in Appendix I, the Scientific
      Authority shall advise the appropriate Management Authority
      of suitable measures to be taken to limit the grant of export
      permits for specimens of that species.

4.   The import of any specimen of a species included in Appendix II
      shall require the prior presentation of either an export per-
      mit or a re-export certificate.

5.   The re-export of any specimen of a species included in
      Appendix II shall require the prior grant and presentation of
      a re-export certificate. A re-export certificate shall only
      be granted when the following conditions have been met:

(a)   a Management Authority of the State of re-export is
       satisfied that the specimen was imported into that State
       in accordance with the provisions of the present
       Convention; and

(b)   a Management Authority of the State of re-export is
       satisfied that any living specimen will be so prepared
       and shipped as to minimize the risk of injury, damage
       to health or cruel treatment.

6.   The introduction from the sea of any specimen of a species
      included in Appendix II shall require the prior grant of a
      certificate from a Management Authority of the State of
      introduction. A certificate shall only be granted when the
      following conditions have been met:

      (a)   a Scientific Authority of the State of introduction
             advises that the introduction will not be detrimental
             to the survival of the species involved; and

      (b)   a Management Authority of the State of introduction is
             satisfied that any living specimen will be so handled
             as to minimize the risk of injury, damage to health or
             cruel treatment.

7.   Certificates referred to in paragraph 6 of this Article may
      be granted on the advice of a Scientific Authority, in
      consultation with other national scientific authorities or,
      when appropriate, international scientific authorities, in
      respect of periods not exceeding one year for total numbers
      of specimens to be introduced in such periods.

ARTICLE V

Regulations of Trade in Specimens of Species included in Appendix III

1.   All trade in specimens of species included in Appendix III
      shall be in accordance with the provisions of this Article.

2.   The export of any specimen of a species included in
      Appendix III from any State which has included that species
      in Appendix III shall require the prior grant and presentation
      of an export permit. An export permit shall only be granted
      when the following conditions have been met:

      (a)   a Management Authority of the State of export is satis-
             fied that the specimen was not obtained in contravention
             of the laws of that State for the protection of fauna
             and flora; and

(b)   a Management Authority of the State of export is
        satisfied that any living specimen will be so prepared
        and shipped as to minimize the risk of injury, damage
        to health or cruel treatment.

3.   The import of any specimen of a species included in
      Appendix III shall require, except in circumstances to which
      paragraph 4 of this Article applies, the prior presentation
      of a certificate of origin and, where the import is from a
      State which has included that species in Appendix III, an
      export permit.

4.   In the case of re-export, a certificate granted by the
      Management Authority of the State of re-export that the
      specimen was processed in that State or is being re-exported
      shall be accepted by the State of import as evidence that the
      provisions of the present Convention have been complied with
      in respect of the specimen concerned.

ARTICLE VI

Permits and Certificates

1.   Permits and certificates granted under the provisions of
      Articles III, IV and V shall be in accordance with the pro-
      visions of this Article.

2.   An export permit shall contain the information specified in
      the model set forth in Appendix IV, and may only be used for
      export within a period of six months from the date on which
      it was granted.

3.   Each permit or certificate shall contain the title of the
      present Convention, the name and any identifying stamp of the
      Management Authority granting it and a control number assigned
      by the Management Authority.

4.   Any copies of a permit or certificate issued by a Management
      Authority shall be clearly marked as copies only and no such
      copy may be used in place of the original, except to the
      extent endorsed thereon.

5.   A separate permit or certificate shall be required for each
      consignment of specimens.

6.   A Management Authority of the State of import of any specimen
      shall cancel and retain the export permit or re-export certi-
      ficate and any corresponding import permit presented in
      respect of the import of that specimen.

7. Where appropriate and feasible a Management Authority may affix
    a mark upon any specimen to assist in identifying the specimen.
    For these purposes "mark" means any indelible imprint, lead
    seal or other suitable means of identifying a specimen, designed
    in such a way as to render its imitation by unauthorized
    persons as difficult as possible.

ARTICLE VII

Exemptions and Other Special Provisions Relating to Trade

1. The provisions of Articles III, IV and V shall not apply to
    the transit or trans-shipping of specimens through or in the
    territory of a Party while the specimens remain in Customs
    control.

2. Where a Management Authority of the State of export or re-
    export is satisfied that a specimen was acquired before the
    provisions of the present Convention applied to that specimen,
    the provisions of Articles III, IV and V shall not apply to
    that specimen where the Management Authority issues a
    certificate to that effect.

3. The provisions of Articles III, IV and V shall not apply to
    specimens that are personal or household effects. This
    exemption shall not apply where:

    (a)   in the case of specimens of a species included in
          Appendix I, they were acquired by the owner outside his
          State of usual residence, and are being imported into
          that State; or

    (b)   in the case of specimens of species included in
          Appendix II:

          (i) they were acquired by the owner outside his State
              of usual residence and in a State where removal
              from the wild occurred;

          (ii) they are being imported into the owner's State of
               usual residence; and

          (iii) the State where removal from the wild occurred
                requires the prior grant of export permits before
                any export of such specimens;

    unless a Management Authority is satisfied that the specimens
    were acquired before the provisions of the present Convention
    applied to such specimens.

4. Specimens of an animal species included in Appendix I bred in captivity for commercial purpose, or of a plant species included in Appendix I artificially propagated for commercial purposes, shall be deemed to be specimens of species included in Appendix II.

5. Where a Management Authority of the State of export is satisfied that any specimen of an animal species was bred in captivity or any specimen of a plant species was artificially propagated, or is a part of such an animal or plant or was derived therefrom, a certificate by that Management Authority to that effect shall be accepted in lieu of any of the permits or certificates required under the provisions of Articles III, IV or V.

6. The provisions of Articles III, IV and V shall not apply to the non-commercial loan, donation or exchange between scientists or scientific institutions registered by a Management Authority of their State, of herbarium specimens, other preserved, dried or embedded museum specimens, and live plant material which carry a label issued or approved by a Management Authority.

7. A Management Authority of any State may waive the requirements of Articles III, IV and V and allow the movement without permits or certificates of specimens which form part of a travelling zoo, circus, menagerie, plant exhibition or other travelling exhibition provided that:

    (a) the exporter or importer registers full details of such specimens with that Management Authority;

    (b) the specimens are in either of the categories specified in paragraphs 2 or 5 of this Article; and

    (c) the Management Authority is satisfied that any living specimen will be so transported and cared for as to minimize the risk of injury, damage to health or cruel treatment.

ARTICLE VIII

Measures to be Taken by the Parties

1. The Parties shall take appropriate measures to enforce the provisions of the present Convention and to prohibit trade in specimens in violation thereof. These shall include measures:

(a)   to penalize trade in, or possession of, such specimens, or both; and

(b)   to provide for the confiscation or return to the State of export of such specimens.

2.   In addition to the measures taken under paragraph 1 of this Article, a Party may, when it deems it necessary, provide for any method of internal reimbursement for expenses incurred as a result of the confiscation of a specimen traded in violation of the measures taken in the application of the provisions of the present Convention.

3.   As far as possible, the Parties shall ensure that specimens shall pass through any formalities required for trade with a minimum of delay. To facilitate such passage, a Party may designate ports of exit and ports of entry at which specimens must be presented for clearance. The Parties shall ensure further that all living specimens, during any period of transit, holding or shipment, are properly cared for so long as to minimize the risk of injury, damage to health or cruel treatment.

4.   Where a living specimen is confiscated as a result of measures referred to in paragraph 1 of this Article:

(a)   the specimen shall be entrusted to a Management Authority of the State of confiscation;

(b)   the Management Authority shall, after consultation with the State of export, return the specimen to that State at the expense of that State, or to a rescue centre or such other place as the Management Authority deems appropriate and consistent with the purposes of the present Convention; and

(c)   the Management Authority may obtain the advice of a Scientific Authority, or may, whenever it considers it desirable, consult the Secretariat in order to facilitate the decision under subparagraph (b) of this paragraph, including the choice of a rescue centre or other place.

5.   A rescue centre as referred to in paragraph 4 of the Article means an institution designated by a Management Authority to look after the welfare of living specimens, particularly those that have been confiscated.

6.   Each Party shall maintain records of trade in specimens of species included in Appendices I, II and III which shall cover:

(a)   the names and addresses of exporters and importers; and

(b)   the number and type of permits and certificates granted;
       the States with which such trade occurred; the numbers
       or quantities and types of specimens, names of species
       as included in Appendices I, II and III and, where
       applicable, the size and sex of the specimens in question.

7.   Each Party shall prepare periodic reports on its implementation
      of the present Convention and shall transmit to the
      Secretariat:

      (a)   an annual report containing a summary of the information
             specified in subparagraph (b) of paragraph 6 of this
             Article; and

      (b)   a biennial report on legislative, regulatory and
             administrative measures taken to enforce the provisions
             of the present Convention.

8.   The information referred to in paragraph 7 of this Article
      shall be available to the public where this is not inconsis-
      tent with the law of the Party concerned.

ARTICLE IX

Management and Scientific Authorities

1.   Each Party shall designate for the purpose of the present
      Convention:

      (a)   one or more Management Authorities competent to grant
             permits or certificates on behalf of that Party; and

      (b)   one or more Scientific Authorities.

2.   A State depositing an instrument of ratification, acceptance,
      approval or accession shall at that time inform the Depositary
      Government of the name and address of the Management Authority
      authorized to communicate with other Parties and with the
      Secretariat.

3.   Any changes in the designations or authorizations under the
      provisions of this Article shall be communicated by the Party
      concerned to the Secretariat for transmission to all other
      Parties.

4.   Any Management Authority referred to in paragraph 2 of this
        Article shall if so requested by the Secretariat or the
        Management Authority of another Party, communicate to it
        impression of stamps, seals or other devices used to
        authenticate permits or certificates.

ARTICLE X

Trade with States not Party to the Convention

        Where export or re-export is to, or import is from, a State
        not a party to the present Convention, comparable documenta-
        tion issued by the competent authorities in that State which
        substantially conforms with the requirements of the present
        Convention for permits and certificates may be accepted in
        lieu thereof by any Party.

ARTICLE XI

Conference of the Parties

1.   The Secretariat shall call a meeting of the Conference of the
        Parties not later than two years after the entry into force
        of the present Convention.

2.   Thereafter the Secretariat shall convene regular meetings at
        least once every two years, unless the Conference decides
        otherwise, and extraordinary meetings at any time on the
        written request of at least one-third of the Parties.

3.   At meetings, whether regular or extraordinary, the Parties
        shall review the implementation of the present Convention
        and may:

        (a)   make such provision as may be necessary to enable the
                Secretariat to carry out its duties;

        (b)   consider and adopt amendments to Appendices I and II in
                accordance with Article XV;

        (c)   review the progress made towards the restoration and
                conservation of the species included in Appendices I,
                II and III;

        (d)   receive and consider any reports presented by the
                Secretariat or by any Party; and

(e)   where appropriate, make recommendations for improving the effectiveness of the present Convention.

4.   At each regular meeting, the Parties may determine the time and venue of the next regular meeting to be held in accordance with the provisions of paragraph 2 of this Article.

5.   At any meeting, the Parties may determine and adopt rules of procedure for the meeting.

6.   The United Nations, its Specialized Agencies and the International Atomic Energy Agency, as well as any State not a Party to the present Convention, may be represented at meetings of the Conference by observers, who shall have the right to participate but not to vote.

7.   Any body or agency technically qualified in protection, conservation or management of wild fauna and flora, in the following categories, which has informed the Secretariat of its desire to be represented at meetings of the Conference by observers, shall be admitted unless at least one-third of the Parties present object:

(a)   international agencies or bodies, either governmental or non-governmental, and national governmental agencies and bodies; and

(b)   national non-governmental agencies or bodies which have been approved for this purpose by the State in which they are located.

Once admitted, these observers shall have the right to participate but not to vote.

ARTICLE XII

The Secretariat

1.   Upon entry into force of the present Convention, a Secretariat shall be provided by the Executive Director of the United Nations Environment Programme. To the extent and in the manner he considers appropriate, he may be assisted by suitable inter-governmental or non-governmental, international or national agencies and bodies technically qualified in protection, conservation and management of wild fauna and flora.

2.   The functions of the Secretariat shall be:

(a)   to arrange for and service meetings of the Parties;

(b)   to perform the functions entrusted to it under the pro-
       visions of Articles XV and XVI of the present
       Convention;

(c)   to undertake scientific and technical studies in
       accordance with programmes authorized by the Conference
       of the Parties as will contribute to the implementation
       of the present Convention, including studies concerning
       standards for appropriate preparation and shipment of
       living specimens and the means of identifying specimens;

(d)   to study the reports of Parties and to request from
       Parties such further information with respect thereto
       as it deems necessary to ensure implementation of the
       present Convention;

(e)   to invite the attention of the Parties to any matter
       pertaining to the aims of the present Convention;

(f)   to publish periodically and distribute to the Parties
       current editions of Appendices I, II and III together
       with any information which will facilitate identifica-
       tion of specimens of species included in those
       Appendices.

(g)   to prepare annual reports to the Parties on its work
       and on the implementation of the present Convention and
       such other reports as meetings of the Parties may
       request;

(h)   to make recommendations for the implementation of the
       aims and provisions of the present Convention, including
       the exchange of information of a scientific or technical
       nature; and

(i)   to perform any other function as may be entrusted to it
       by the Parties.

ARTICLE XIII

International Measures

1.   When the Secretariat in the light of information received is
      satisfied that any species included in Appendices I or II is
      being affected adversely by trade in specimens of that species
      or that the provisions of the present Convention are not being

effectively implemented, it shall communicate such information to the authorized Management Authority of the Party or Parties concerned.

2. When any Party receives a communication as indicated in paragraph 1 of this Article, it shall, as soon as possible, inform the Secretariat of any relevant facts insofar as its laws permit and, where appropriate, propose remedial action. Where the Party considers that an inquiry is desirable, such inquiry may be carried out by one or more persons expressly authorized by the Party.

3. The information provided by the Party or resulting from any inquiry as specified in paragraph 2 of this Article shall be reviewed by the next Conference of the Parties which may make whatever recommendation it deems appropriate.

ARTICLE XIV

Effect on Domestic Legislation and International Conventions

1. The provisions of the present Convention shall in no way affect the right of Parties to adopt:

    (a) stricter domestic measures regarding the conditions for trade, taking possession or transport of specimens of species included in Appendices I, II and III, or the complete prohibition thereof; or

    (b) domestic measures restricting or prohibiting trade, taking possession, or transport of species not included in Appendices I, II and III.

2. The provisions of the present Convention shall in no way affect the provisions of any domestic measures or the obligations of Parties deriving from any treaty, convention, or international agreement relating to other aspects of trade, taking possession, or transport of specimens which is in force or subsequently may enter into force for any Party including any measure pertaining to the Customs, public health, veterinary or plant quarantine fields.

3. The provisions of the present Convention shall in no way affect the provisions of, or the obligations deriving from, any treaty, convention or international agreement concluded or which may be concluded between States creating a union or regional trade agreement establishing or maintaining a common external customs control and removing customs control between

the parties thereto insofar as they relate to trade among the States members of that union agreement.

4.   A State Party to the present Convention, which is also a party to any other treaty, convention or international agreement which is in force at the time of the coming into force of the present Convention and under the provisions of which protection is afforded to marine species included in Appendix II, shall be relieved of the obligation imposed on it under the provisions of the present Convention with respect to trade in specimens of species included in Appendix II that are taken by ships registered in that State and in accordance with the provisions of such other treaty, convention or international agreement.

5.   Notwithstanding the provisions of Articles III, IV and V, any export of a specimen taken in accordance with paragraph 4 of this Article shall only require a certificate from a Management Authority of the State of introduction to the effect that the specimen was taken in accordance with the provisions of the other treaty, convention or international agreement in question.

6.   Nothing in the present Convention shall prejudice the codification and development of the law of the sea by the United Nations Conference on the Law of the Sea convened pursuant to Resolution 2750 C (XXV) of the General Assembly of the United Nations nor the present or future claims and legal views of any State concerning the law of the sea and the nature and extent of coastal and flag State jurisdiction.

ARTICLE XV

Amendments to Appendices I and II

1.   The following provisions shall apply in relation to amendments to Appendices I and II at meetings of the Conference of the Parties:

      (a)   Any Party may propose an amendment to Appendix I or II for consideration at the next meeting. The text of the proposed amendment shall be communicated to the Secretariat at least 150 days before the meeting. The Secretariat shall consult the other Parties and interested bodies on the amendment in accordance with the provisions of subparagraph (b) and (c) of paragraph 2 of this Article and shall communicate the response to all Parties not later than 30 days before the meeting.

(b)   Amendments shall be adopted by a two-thirds majority of
         Parties present and voting. For these purposes "Parties
         present and voting" means Parties present and casting an
         affirmative or negative vote. Parties abstaining from
         voting shall not be counted among the two-thirds required
         for adopting an amendment.

(c)   Amendments adopted at a meeting shall enter into force
         90 days after that meeting for all Parties except those
         which make a reservation in accordance with paragraph 3
         of this Article.

2.   The following provisions shall apply in relation to amendments
     to Appendices I and II between meetings of the Conference of
     the Parties:

(a)   Any Party may propose an amendment to Appendix I or II
         for consideration between meetings by the postal pro-
         cedures set forth in this paragraph.

(b)   For marine species, the Secretariat shall, upon receiving
         the text of the proposed amendment, immediately com-
         municate it to the Parties. It shall also consult inter-
         governmental bodies having a function in relation to
         those species especially with a view to obtaining
         scientific data these bodies may be able to provide and
         to ensuring co-ordination with any conservation measures
         enforced by such bodies. The Secretariat shall com-
         municate the views expressed and data provided by these
         bodies and its own findings and recommendations to the
         Parties as soon as possible.

(c)   For species other than marine species, the Secretariat
         shall, upon receiving the text of the proposed amendment,
         immediately communicate it to the Parties, and as soon
         as possible thereafter, its own recommendations.

(d)   Any Party may, within 60 days of the date on which the
         Secretariat communicated its recommendations to the
         Parties under subparagraphs (b) and (c) of this para-
         graph, transmit to the Secretariat any comments on the
         proposed amendment together with any relevant scientific
         data and information.

(e)   The Secretariat shall communicate the replies received
         together with its own recommendations to the Parties as
         soon as possible.

(f)   If no objection to the proposed amendment is received
         by the Secretariat within 30 days of the date the

replies and recommendations were communicated under the provisions of subparagraph (e) of this paragraph, the amendment shall enter into force 90 days later for all Parties except those which make a reservation in accordance with paragraph 3 of this Article.

(g)   If an objection by any Party is received by the Secretariat, the proposed amendment shall be submitted to a postal vote in accordance with the provisions of subparagraphs (h), (i) and (j) of this paragraph.

(h)   The Secretariat shall notify the Parties that notification of objection has been received.

(i)   Unless the Secretariat receives the votes for, against or in abstention from at least one-half of the Parties within 60 days of the date of notification under subparagraph (h) of this paragraph, the proposed amendment shall be referred to the next meeting of the Conference for further consideration.

(j)   Provided that votes are received from one-half of the Parties, the amendment shall be adopted by a two-thirds majority of Parties casting an affirmative or negative vote.

(k)   The Secretariat shall notify all Parties of the result of the vote.

(l)   If the proposed amendment is adopted it shall enter into force 90 days after the date of the notification by the Secretariat of its acceptance for all Parties except those which make a reservation in accordance with paragraph 3 of this Article.

3.   During the period of 90 days provided for by subparagraph (c) of paragraph 1 or subparagraph (l) of paragraph 2 of this Article any Party may by notification in writing to the Depositary Government make a reservation with respect to the amendment. Until such reservation is withdrawn the Party shall be treated as a State not a Party to the present Convention with respect to trade in the species concerned.

ARTICLE XVI

Appendix III and Amendments thereto

1.   Any Party may at any time submit to the Secretariat a list of

species which it identifies as being subject to regulation
within its jurisdiction for the purpose mentioned in paragraph 3 of Article II, Appendix III shall include the names
of the Parties submitting the species for inclusion therein,
the scientific names of the species so submitted, and any
parts or derivatives of the animals or plants concerned that
are specified in relation to the species for the purposes of
subparagraph (b) of Article I.

2. Each list submitted under the provisions of paragraph 1 of
    this Article shall be communicated to the Parties by the
    Secretariat as soon as possible after receiving it. The list
    shall take effect as part of Appendix III 90 days after the
    date of such communication. At any time after the communica-
    tion of such list, any Party may by notification in writing
    to the Depositary Government enter a reservation with respect
    to any species or any parts or derivatives, and until such
    reservation is withdrawn, the State shall be treated as a
    State not a Party to the present Convention with respect to
    trade in the species or part or derivative concerned.

3. A Party which has submitted a species for inclusion in
    Appendix III may withdraw it at any time by notification to
    the Secretariat which shall communicate the withdrawal to all
    Parties. The withdrawal shall take effect 30 days after the
    date of such communication.

4. Any Party submitting a list under the provisions of paragraph 1
    of this Article shall submit to the Secretariat a copy of all
    domestic laws and regulations applicable to the protection of
    such species, together with any interpretations which the
    Party may deem appropriate or the Secretariat may request.
    The Party shall, for as long as the species in question is
    included in Appendix III, submit any amendment of such laws
    and regulations or any new interpretations as they are adopted.

ARTICLE XVII

Amendment of the Convention

1. An extraordinary meeting of the Conference of the Parties
    shall be convened by the Secretariat on the written request
    of at least one-third of the Parties to consider and adopt
    amendments to the present Convention. Such amendments shall
    be adopted by a two-thirds majority of Parties present and
    voting. For these purposes "Parties present and voting" means
    Parties present and casting an affirmative or negative vote.
    Parties abstaining from voting shall not be counted among the
    two-thirds required for adopting an amendment.

2.   The text of any proposed amendment shall be communicated by
      the Secretariat to all Parties at least 90 days before the
      meeting.

3.   An amendment shall enter into force for the Parties which
      have accepted it 60 days after two-thirds of the Parties have
      deposited an instrument of acceptance of the amendment with
      the Depositary Government. Thereafter, the amendment shall
      enter into force for any other Party 60 days after that Party
      deposits its instrument of acceptance of the amendment.

ARTICLE XVIII

Resolution of Disputes

1.   Any dispute which may arise between two or more Parties with
      respect to the interpretation or application of the provisions
      of the present Convention shall be subject to negotiation
      between the Parties involved in the dispute.

2.   If the dispute cannot be resolved in accordance with para-
      graph 1 of this Article, the Parties may, by mutual consent,
      submit the dispute to arbitration, in particular that of the
      Permanent Court of Arbitration at The Hague and the Parties
      submitting the dispute shall be bound by the arbitral
      decision.

ARTICLE XIX

Signature

      The present Convention shall be open for signature at
      Washington until 30th April 1973 and thereafter at Berne
      until 31st December 1974.

ARTICLE XX

Ratification, Acceptance, Approval

      The present Convention shall be subject to ratification,
      acceptance or approval. Instruments of ratification, accep-
      tance or approval shall be deposited with the Government of
      the Swiss Confederation which shall be the Depositary Govern-
      ment.

ARTICLE XXI

Accession

The present Convention shall be open indefinitely for accession. Instruments of accession shall be deposited with the Depositary Government.

ARTICLE XXII

Entry into Force

1. The present Convention shall enter into force 90 days after the date of deposit of the tenth instrument of ratification, acceptance, approval or accession, with the Depositary Government.

2. For each State which ratifies, accepts or approves the present Convention or accedes thereto after the deposit of the tenth instrument of ratification, acceptance, approval or accession, the present Convention shall enter into force 90 days after the deposit by such State of its instrument of ratification, acceptance, approval or accession.

ARTICLE XXIII

Reservations

1. The provisions of the present Convention shall not be subject to general reservations. Specific reservations may be entered in accordance with the provisions of this Article and Articles XV and XVI.

2. Any State may, on depositing its instrument of ratification, acceptance, approval or accession, enter a specific reservation with regard to:

    (a)   any species included in Appendix I, II or III; or

    (b)   any parts or derivatives specified in relation to a species included in Appendix III.

3. Until a Party withdraws its reservation entered under the provisions of this Article, it shall be treated as a State not a Party to the Present Convention with respect to trade

in the particular species or parts or derivatives specified
in such reservation.

ARTICLE XXIV

Denunciation

Any Party may denounce the present Convention by written
notification to the Depositary Government at any time. The
denunciation shall take effect twelve months after the
Depositary Government has received the notification.

ARTICLE XXV

Depositary

1.   The original of the present Convention, in the Chinese,
     English, French, Russian and Spanish languages, each version
     being equally authentic, shall be deposited with the
     Depositary Government, which shall transmit certified copies
     thereof to all States that have signed it or deposited
     instruments of accession to it.

2.   The Depositary Government shall inform all signatory and
     acceding States and the Secretariat of signatures, deposit of
     instruments of ratification, acceptance, approval or accession,
     entry into force of the present Convention, amendments thereto,
     entry and withdrawal of reservations and notifications of
     denunciation.

3.   As soon as the present Convention enters into force, a
     certified copy thereof shall be transmitted by the Depositary
     Government to the Secretariat of the United Nations for
     registration and publication in accordance with Article 102
     of the Charter of the United Nations.

     In witness whereof the undersigned Plenipotentiaries, being
     duly authorized to that effect, have signed the present
     Convention.

     Done at Washington this third day of March, One Thousand Nine
     Hundred and Seventy-three.

APPENDIX I

Interpretation:

1. Species included in this Appendix are referred to:

   (a) by the name of the species; or

   (b) as being all of the species included in a higher
        taxon or designated part thereof.

2. The abbreviation "spp." is used to denote all species of
   a higher taxon.

3. Other references to taxa higher than species are for the
   purposes of indormation or classification only.

ARACEAE
  Alocasia sanderana
  Alocasia zebrina
CARYOCARACEAE
  Caryocar costaricense
CARYOPHYLLACEAE
  Gymnocarpos przewalskii
  Melandrium mongolicum
  Silene mongolica
  Stellaria pulvinata
CUPRESSACEAE
  Fitzroya cupressoides
  Pilgerodendron uviferum
CYCADACEAE
  Encephalartos spp.
  Microcycas calocoma
  Stangeria eriopus
GENTIANACEAE
  Prepusa hookerana
HUMIRIACEAE
  Vantanea barbourii
JUGLANDACEAE
  Engelhardtia pterocarpa
LEGUMINOSAE
  Ammopiptanthus mongolicus
  Cynometra hemitomophylla
  Platymiscium pleiostachyum
  Tachigalia versicolor
LILIACEAE
  Aloe albida
  Aloe pillansii
  Aloe polyphylla
  Aloe thorncroftii
  Aloe vossii

MELASTOMATACEAE
  Lavoisiera itambana
MELIACEAE
  Guarea longipetiola
MORACEAE
  Batocarpus costaricensis
ORCHIDACEAE
  Cattleya skinneri
  Cattleya trianae
  Didiciea cunninghamii
  Laelia jongheana
  Laelia lobata
  Lycaste virginalis var. alba
  Peristeria elata
PINACEAE
  Abies guatemalensis
  Abies nebrodensis
PODOCARPACEAE
  Podocarpus costalis
  Podocarpus parlatorei
PROTEACEAE
  Orothamnus zeyheri
  Protea odorata
RUBIACEAE
  Balmea stormae
SAXIFRAGACEAE (GROSSULARIACEAE)
  Ribes sardoum
ULMACEAE
  Celtis aetnensis
WELWITSCHIACEAE
  Welwitschia bainesii
ZINGIBERACEAE
  Hedychium philippinense

APPENDIX II

Interpretation:

1. Species included in this Appendix are referred to:

    (a) by the name of the species; or

    (b) as being all of the species included in a higher
         taxon or designated part thereof.

2. The abbreviation "spp." is used to denote all the species
    of a higher taxon.

3. Other references to taxa higher than species are for the
    purposes of information or classification only.

4. The symbol (*) followed by a number placed against the
    name of a species or higher taxon designates parts or
    derivatives which are specified in relation thereto for
    the purposes of the present Convention as follows:

        * 1   designates root
        * 2   designates timber
        * 3   designates trunks

5. The symbol (-) followed by a number placed against the
    name of a species or higher taxon indicates the exclusion
    from that species or taxon of designated geographically
    separate populations, subspecies, species or groups of
    species as follows:

        - 4   Species which are not succulents

6. The symbol (+) followed by a number placed against the
    name of a species or higher taxon denotes that only
    designated geographically separate populations, subspecies
    or species of that species or taxon are included in this
    Appendix as follows:

        + 5   All species of the family in the Americas

APOCYNACEAE	Pachypodium spp.
ARALIACEAE	Panax quinquefolius * 1
ARAUCARIACEAE	Araucaria araucana * 2
CACTACEAE	Cactaceae spp. + 5 Rhipsalis spp.
COMPOSITAE	Saussurea lappa * 1
CYATHEACEAE	Cyathea (Hemitelia) capensis * 3 Cyathea dregei * 3 Cyathea mexicana * 3 Cyathea (Alsophila) salvinii * 3
DIOSCOREACEAE	Dioscorea deltoidea * 1
EUPHORBIACEAE	Euphorbia spp. - 4
FAGACEAE	Quercus copeyensis * 2
LEGUMINOSAE	Thermopsis mongolica
LILIACEAE	Aloe spp.
MELIACEAE	Swietenia humilis * 2
ORCHIDACEAE	Spp.
PALMAE	Areca ipot Phoenix hanceana var. philippinensis Zalacca clemensiana
PORTULACACEAE	Anacampseros spp.
PRIMULACEAE	Cyclamen spp.
SOLANACEAE	Solanum sylvestre
STERCULIACEAE	Basiloxylon excelsum * 2
VERBENACEAE	Caryopteris mongolica
ZYGOPHYLLACEAE	Guaiacum sanctum * 2

DISCUSSION SECTION X INTERNATIONAL CO-OPERATION AND LEGISLATION

Professor J Heslop-Harrison (Kew) thanked Sir Otto Frankel for stressing the ethical aspects of conservation which are of great importance and said that we must not under-estimate the sympathy in this area from a generally uninformed public. Quick results must be achieved by agreeing on practical attitudes and practical reasons for our conservation work which will convince politicians of its importance.

He announced that the NATO Eco-Sciences Panel is sponsoring a conference on the rehabilitation of damaged ecosystems which is to be held in Iceland during 1976.

The International Board of Genetic Research Centres covers the conservation of economically important crop plants and those considered to be of potential importance. The limitations on what they think should be conserved sometimes narrows the range of species covered. Large scale chemical surveys of plants have revealed many plants of potential use and where the conservation of these was not being undertaken by the Genetic Resource Centres they should be covered by botanic gardens.

He concluded by quoting a Maoist saying that he had recently heard which demonstrated the contrast between the traditional Western and Eastern ethic with regard to nature: "The natural world exists for the benefit of man, therefore man must look after the natural world very well indeed".

Dr Melville (Kew) agreed with Sir Otto's statements that plants and animals have their own right to evolution, but was concerned about his emphasis on crop plants and their wild relatives which suggested that the remainder are not important. The majority of crop plants are from very limited sources and the floras of many parts of the world remain untapped. He cited Cordeauxia edulis, an edible plant in Somalia which has been almost exterminated, but has been saved by the World Wildlife Fund. It is particularly valuable because it is a legume suited to very low rainfall areas. The Macadamia nut is one of the few examples of an edible nut from Australia and there are probably more which might be developed into food plants. These are strong reasons for conserving a wider range of material.

305

Sir Otto Frankel (Canberra) said that the meeting had refrained from discussing reserves which are the rightful depositories of wild biota, since the conference was emphasising the role of botanic gardens in conservation. Conservation of certain plants which might be a resource useful to the chemical industry might be desirable but chemical surveys have been continuing for decades without much return. He doubted whether there would be much gain in useful food plants from similar surveys since the number of forms already in use has recently been diminishing. The reserves of fruit present, for example, in Latin America and in South East Asia have not been considered, and these are probably in danger.

Dr Hall (Cape Town) asked how we can supply this information to the public and administrators in order to allow botanic gardens to act in a positive way.

Sir Otto Frankel (Canberra) replied that the Biosphere concept which is now being promoted is directed mainly at the developing countries of Africa, where much is threatened and much needs to be conserved. Progress is being made and some of the African National Parks and Reserves will be upgraded due to the Biosphere programme.

Dr Williams (Birmingham) said we must be careful of distinctions drawn between cultivated plants and wild species. Many crop plants are only partially domesticated and are sometimes of very local significance in developing countries. Botanic gardens may have a role to play in the conservation of these plants.

Dr Keay (Royal Society, London) said that in order to gain the support of the public, administrators and others controlling funds there is a need for some immediate and effective action. Priority should be given to the conservation of.wild and domesticated species using existing resources in order to show that a worthwhile task is being successfully initiated.

Professor Zohary (Jerusalem) pointed out the importance of conserving fruit trees with sampling from a sufficient number and range to give an adequate gene pool.

Sir Otto Frankel (Canberra) agreed that fruit species are of concern at wild or semi-wild levels. Collections of some wild material have been made but special plantations would be needed if extensive collections of cultivars were required, but he doubted the necessity of this.

Iran had been the cradle of the development of many fruit trees grown popularly in Western Europe, and yet there is not a single establishment for the conservation of local fruit trees in that country. In Turkey it was considered that home gardens

represented the best sources of material for conservation. The
International Board for Plant Genetic Resources had set up a
committee of the five principal asiatic countries involved to
consider the conservation of fruit species.

Discussion on Report of IABG Meeting

Professor Hawkes (Birmingham) asked Mr Henderson to what
extent the discussions at the Moscow meeting of the IABG centred
on conserving material in botanic gardens and the method of
sampling.

Mr Henderson (Edinburgh) replied that the gene pool situation
was not discussed. It was primarily concerned with a survey of
ideas on conservation and what botanic gardens in the USSR are
doing. He pointed out that in most respects the Moscow IABG
meeting and the present conference are complementary.

Dr Wijnands (Amsterdam) asked about the future plans of IABG.

Mr Henderson (Edinburgh) said that the next main meeting would
be held in Sydney at the International Botanical Congress in 1981.
A new council was elected at Moscow.

Discussion on International Co-operation and Legislation

Professor Cook (Zurich) asked Mr Lucas about publicising to
taxonomists the forms used by the Threatened Plants Committee for
gathering information.

Mr Lucas (Kew) replied that this was a phase which he hoped
to have in operation soon.

Professor Ehrendorfer (Vienna) pointed out that a major part
of the problem of conservation was in the tropics and that there
was too little co-operation between temperate and tropical
institutions. He asked Mr Lucas whether the IUCN could organise
this co-operation. He also asked whether they could buy areas
which were still cheap to be used as reserves, or establish field
stations which could later be turned into reserves. He would
welcome the IUCN seeing this as an important goal.

Mr Lucas (Kew) replied that if the Threatened Plants Committee
was given the task it would be glad to do it. If botanic gardens
with access to money and with interest in conserving particular
areas contacted the TPC he would gladly make the necessary contacts.
He pointed out that the need was for small plant reserves such as
the Mutomo Hill Reserve in Kenya. However, there was a problem

that the TPC must not be seen to be politically involved and that projects must be made mutually beneficial. Another snag was that protective measures could become difficult to enforce when the area surrounding a reserve became heavily overgrazed.

Professor J Heslop-Harrison (Kew) stressed that IUCN, through the TPC was not intending to replace existing organisations but to bring such bodies into co-operation to prevent overlap. Referring to Professor Cook's question, he said that the major priority was to produce amicable co-operative arrangements between institutions with the knowledge and those with the need. Many of these may involve relationships of a political nature and call for careful handling of delicate situations.

Resolutions

RESOLUTIONS

1. This Conference, conscious that the rich tropical floras of
    the world are now in great hazard, (1) urges that a strong
    network of nature reserves and conservation-orientated gardens
    should be established throughout the tropics both through the
    strengthening and development of existing foundations and
    through the creation of new ones where the need exists;
    (2) recommends that institutions in temperate countries
    should offer all possible help in this programme through
    technical aid, training and the secondment of personnel; and
    (3) urges that this aim should be pursued through the
    International Union for Conservation of Nature and Natural
    Resources to ensure good co-ordination and proper understand-
    ing of the importance of the work for the tropical countries
    themselves and for the whole of mankind.

2. This Conference urges that special attention be given to the
    Conservation of Threatened Floras particularly of Islands and
    those parts of the world with Mediterranean or similar
    climates since both are often inhabited by very large numbers
    of narrowly endemic species of plants endangered by human
    activities.

3. This Conference recommends that institutions maintaining
    plant collections (including seed collections) for conservation
    purposes should, in general, give priority to their local
    flora, so as (1) to benefit from local taxonomic ecological,
    physiological and other pertinent specialist knowledge;
    (2) to reduce the need to simulate remote climates with the
    attendant costs and dependence on man-generated energy;
    (3) to be able to offer from direct experience information
    and advice concerning field conservation in the country of
    the institution, and (4) to provide a basis from which public
    interest and pride in the indigenous flora can be developed
    through display and education services.

4. This Conference urges all Governments to ratify the
    "Convention on International Trade in Endangered Species of
    Wild Fauna and Flora" as soon as possible.

5. This Conference recommends that, wherever possible, all living
    plant collections grown for conservation purposes should also

be stored in the form of seeds under appropriate conditions for long-term conservation.

6. This Conference urges that the propagation of rare and endangered species, including research into appropriate techniques, should be actively pursued by Botanic Gardens and other bodies maintaining living plant collections, and that such activities should be financially supported where necessary by Conservation, or other appropriate Organisations. Special attention should be given to economic plants and their wild relatives and to plants which are or might be commercially used.

7. This Conference urges that whenever threatened plants are taken into cultivation, this be done by means of seed and/or cuttings whenever possible so as not to deplete the wild populations.

8. This Conference, aware of the urgent need for scientifically verified lists of threatened species on a world scale, calls for the full support for the work of the IUCN Threatened Plants Committee in compiling such lists, and urges the task of propagating stocks of species on institutions maintaining living plant collections.

9. This Conference calls for the widest publicity to its full deliberations to be given in all appropriate quarters, and urges that the resolutions should be made available separately for this purpose with the minimum delay.

10. This Conference, being acutely aware of the urgency and complexity of many problems which have been raised during the sessions, urges the desirability of continued study and exchange of information, and the setting up of working parties to continue the study of outstanding issues, e.g.

    1. listing of collections, documentation and dissemination of information

    2. commercial use of wild species

    3. preparation of codes of practise

    4. publicity

    5. relationship between institutions maintaining living plant collections and organisations concerned with nature conservation

    6. compilation of a short list of rare and endangered plants of high scientific importance to be commended to botanic gardens to bring them into cultivation.

LIST OF DELEGATES AND PARTICIPANTS

Mrs M T F de Almeida, Jardin Botanico, Instituto Botânico,
     Universidade de Coimbra, PORTUGAL

Mr J Apel, Technical Leader, Institut für Allgemeine Botanik und
     Botanischer Garten, D-2000 Hamburg, Heston 10,
     FEDERAL REPUBLIC OF GERMANY

Miss R C R Angel, Museums Division, Royal Botanic Gardens, Kew,
     Richmond, Surrey TW9 3AB, ENGLAND

*Professor E S Ayensu, Chairman of Botany, Department of Botany,
     Smithsonian Institution, Washington DC 20560, U.S.A.

Professor T Baytop, Faculty of Pharmacy, University of Istanbul,
     Istanbul, TURKEY

Mr R I Beyer, Deputy Curator, Royal Botanic Gardens, Kew, Richmond,
     Surrey TW9 3AB, ENGLAND

Dr F de Beaufort, Chef du Bureau de la faune et de la flore,
     Direction de la protection de la nature, Ministère de la
     Qualité de la L'l (Environment), 14 Boulevard de Général Leclerc
     92521 Neuilly-sur-Seine, FRANCE

Mr L Bisset, Assistant Curator, Royal Botanic Garden, Edinburgh
     EH3 5LR, SCOTLAND

*Professor T Böcher, Director, Institute of Plant Anatomy and
     Cytology, Copenhagen University, Solvgade 83, Copenhagen
     DENMARK

Dr L Borgen, Curator, Botanical Garden, University of Oslo,
     Trondheimsun 23B, Oslo 5, NORWAY

Dr J van Borssum Waalkes, Curator of Collections, State University
     of Groningen, Biological Centre, Hortus De Wolf, Kerklaan 30,
     Haren (Gr.) NETHERLANDS

Mr A Brady, Director, National Botanic Gardens, Glasnevin,
     Dublin 9, EIRE

Dr D Bramwell, Director, Jardin Botanico 'Viera y Clavijo',
     Tafira Alta, Las Palmas, GRAN CANARIA

Dr P Brandham, Cytology, Royal Botanic Gardens, Kew, Richmond,
     Surrey TW9 3AB, ENGLAND

Mr J P M Brenan, Deputy Director, Royal Botanic Gardens, Kew,
     Richmond, Surrey TW9 3AE, ENGLAND

*Mr C Brickell, Director, Royal Horticultural Society's Garden
     Wisley, Ripley, Woking, Surrey, ENGLAND

Mrs M Briggs, Botanical Society of the British Isles, White Cottage
     Slinfold, Horsham, Sussex, ENGLAND

Miss C Brighton, Cytology, Royal Botanic Gardens, Kew, Richmond,
     Surrey TW9 3AB, ENGLAND

*Mr B F Bruinsma, Curator, University Botanic Garden, Nonnesteeg 3,
     Leiden, NETHERLANDS

Mr G E Brown, Assistant Curator, Arboretum, Royal Botanic Gardens,
     Kew, Richmond, Surrey TW9 3AB, ENGLAND

*Dr G Budowski, Director General, IUCN, 1110 Morges, SWITZERLAND

Mr J K Burras, Superintendent, University Botanic Garden, Oxford,
     ENGLAND

Miss A Chabert, Living Collections Division, Royal Botanic Gardens,
     Kew, Richmond, Surrey TW9 3AB, ENGLAND

Dr G and Dr C Ciuffi, Orto Botanico, Firenze, ITALY

*Professor C D K Cook, Director, Botanischer Garten der Universität
     Zürich, Pelikan Strasse 40, 8001 Zürich, SWITZERLAND

*Dr V E Corona, Director, Jardin Botanico, Universidad Nacional
     Autonoma de Mexico, Mexico City 20 DF, MEXICO

Mr T Crawford, Curator, National Botanic Gardens, Glasnevin,
     Dublin, EIRE

Mr G Crompton, Species Recorder, University Botanic Garden,
     1 Brookside, Cambridge, ENGLAND

Mr T S Crosby, Curator of Botany Experimental Gardens, Department
    of Plant Sciences, The University, Leeds LS2 9JT, ENGLAND

Mr R E Crowther, Forestry Commission, Alice Holt Research Station,
    Farnham, Surrey, ENGLAND

*Dr J Cullen, Royal Botanic Garden, Edinburgh EH3 5LR, SCOTLAND

Mr E W Curtis, Curator, Botanic Gardens, Glasgow G12 OUE, SCOTLAND

Professor H Demiriz, Director, University of Diyarbakir,
    Faculty of Science, Section of Botany, Diyarbakir, TURKEY

Mr J Donald, Ashington Botanical Trust, Holly Gate, Ashington,
    Sussex RH20 3BA ENGLAND

Dr A Doroszewka, Director, Botanical Garden of Warsaw University,
    Al Ujazdowskie 4, 00-478 Warsawa, POLAND

Professor F Ehrendorfer, Director, Botanischer Garten, Universitat
    Wien, A1030 Wien III, Rennweg 14, AUSTRIA

Mr C Ellingworth, University of Birmingham, Department of Botany
    Botanic Garden, 56 Edgbaston Park Road, Birmingham B15 2TT
    ENGLAND

Mr C M Erskine, Assistant Curator, Temperate Department,
    Royal Botanic Gardens, Kew, Richmond, Surrey TW9 3AB, ENGLAND

*Professor Dr K Esser, Director, Ruhr-Universität Bochum Botanischer
    Garten, 4630 Bochum-Querenburg, Buscheystrasse 132,
    Postfach 2148, Bochum, FEDERAL REPUBLIC OF GERMANY

Mr A Evans, Assistant Curator, Royal Botanic Garden,
    Edinburgh EH3 5LR, SCOTLAND

Dr C Farron, Curator of Herbarium and Library, Botanisches
    Institut der Universitat, Schoenbeinstrasse 6, CH 4056 Basel,
    SWITZERLAND

Mr D Field, Herbarium, Royal Botanic Gardens, Kew, Richmond,
    Surrey TW9 3AE, ENGLAND

Dr P A Florschütz, Director, Botanic Garden,
    Botanical Museum and Herbarium, Heidelberglaan 2, Utrecht,
    NETHERLANDS

* Sir Otto Frankel FRS, Senior Research Fellow, CSIRO,
    Division of Plant Industry, PO Box 1600, Canberra City,
    AUSTRALIA

Dr George W Gillett, University of California, Botanical Gardens,
    Riverside, California 92502, U.S.A.

Professor C Gómez-Campo and
    Mrs M Estrella Gómez-Campo (Maestro Laboratorio)
    Department Plant Physiology, Escuela T S Ing. Agrónomos
    Universidad Politecnica, Madrid-3, SPAIN

Mr P S Green, Deputy Keeper, Herbarium, Royal Botanic Gardens,
    Kew, Richmond, Surrey TW9 3AE, ENGLAND

Mr L Guignolini, Technical Manager, Orto Botanico, Firenze, ITALY

*Dr A V Hall, Bolus Herbarium, University of Capetown,
    PO Rondesbosch 7700, Capetown, SOUTH AFRICA

Mr B Halliwell, Assistant Curator, Alpine and Herbaceous Section,
    Royal Botanic Gardens, Kew, Richmond, Surrey TW9 3AB, ENGLAND

Dr K Hare, University of Toronto, 100 St George Street, Toronto,
    Ontario, CANADA

Professor B J Harris, Department of Biological Sciences,
    Ahmadu Bello University, Zaria, NIGERIA

*Professor J G Hawkes, Department of Botany, University of
    Birmingham, PO Box 363 Birmingham B15 2TT, ENGLAND

* Mr Hebb, Horticulturalist, Cary Arboretum of the New York
    Botanical Garden, Box AB, Millbrook, New York, U.S.A.

Professor O Hedberg, Director, and Dr I Hedberg,
    Institute of Systematic Botany, PO Box 571, S-751 21 Uppsala,
    SWEDEN

* Mr D Henderson, Regius Keeper, Royal Botanic Garden,
    Edinburgh EH3 5LR, SCOTLAND

*Professor J Heslop-Harrison, Director, and Dr Y Heslop-Harrison,
    Royal Botanic Gardens, Kew, Richmond, Surrey TW9 3AB, ENGLAND

Mr T M Hewitt, Curator, Ashington Botanical Trust, Holly Gate,
    Ashington, Sussex RH20 3BA, ENGLAND

*Professor V H Heywood, Department of Botany, Plant Science
    Laboratories, The University, Whiteknights, Reading RG6 2AS,
    ENGLAND

Mr H G Hillier, The Hillier Gardens and Arboretum, Ampfield,
     Near Romsey, Hampshire, ENGLAND

Mr J P Hjerting, Institute of Plant Anatomy and Cytology,
     Copenhagen University, Solvgade 83, Copenhagen, DENMARK

Mr P Hoek, Berkendaal 141, Rotterdam Z, NETHERLANDS

*Dr M Hofmann, 775 Konstanz, Buhlenweg 38, FEDERAL REPUBLIC OF
     GERMANY

*Professor W Hondelmann, Forschungsanstalt für Landwirtschaft,
     Braunschweig-Völkenrode, 33 Braunschweig, Bundesalle 50,
     FEDERAL REPUBLIC OF GERMANY

Mr D R Hunt, Herbarium, Royal Botanic Gardens, Kew, Richmond,
     Surrey TW9 3AE, ENGLAND

Professor H Hurka, Botanischer Garten und Botanisches Institut
     der Universitat, D-44 Munster, Schlossgarten 3,
     FEDERAL REPUBLIC OF GERMANY

Mr J Iff, Chef du Jardin, Conservatoire et Jardin Botaniques,
     192 route de Lausanne, 1202 Genève, SWITZERLAND

Dr H S Irwin, President, New York Botanic Garden, Bronx,
     New York 10458, U.S.A.

Dr D W Jeffrey, Acting Director, Botanic Garden, Trinity College,
     Dublin 2, EIRE

Dr K Jones, Keeper of Jodrell Laboratory, Royal Botanic Gardens,
     Kew, Richmond, Surrey TW9 3AB, ENGLAND

Dr R W J Keay, The Royal Society, 6 Carlton House Terrace,
     London SW17 5AG, ENGLAND

Mr B Kirchner, Ruhr-Universität Bochum Botanischer Garten,
     4630 Bochum-Querenburg, Buscheystrasse 132, Postfach 2148,
     Bochum, FEDERAL REPUBLIC OF GERMANY

Professor H Kristinsson, Director of Museum, Akureyri Museum of
     Natural History, PO Box 580, Akureyri, ICELAND

Mr E Lammens, Chef de Travaux, Jardin Botanique National de
     Belgique, Domaine de Bouchout, 1800 Meise, BELGIUM

Mr C R Lancaster, Curator, The Hillier Gardens and Arboretum,
     Ampfield, Near Romsey, Hampshire, ENGLAND

Professor J F Leroy, Museum National d'Histoire Naturelle,
      57 Rue Cuvier, Paris 5, FRANCE

Mr J Y Lesouëf, SE PNB, Faculté des Sciences, Avenue Le Gorgeu,
      29 Brest, FRANCE

*Mr G Ll Lucas, Conservation Unit, Royal Botanic Gardens, Kew,
      Richmond, Surrey TW9 3AE, ENGLAND

Mr J D Main, Superintendent, Northern Horticultural Society,
      Harlow Car Gardens, Harrogate, Yorkshire, ENGLAND

Mr P J Maudsley, Royal Botanic Garden, Edinburgh EH3 5LR, SCOTLAND

Mr R W Mellors, Conservation Propagator, University Botanic Garden,
      1 Brookside, Cambridge, ENGLAND

Dr E A Mennega, Scientific Officer, Botanical Museum and Herbarium,
      Heidelberglaan 2, Utrecht, NETHERLANDS

Dr R Melville, 121 Mortlake Road, Kew, Richmond, Surrey, ENGLAND

Professor H Merxmüller, Director, Botanische Staatssammlung und
      Botanischer Garten, D-8 München 19, Menzingerstrasse 63,
      FEDERAL REPUBLIC OF GERMANY

Professor J Miège, Director, Conservatoire et Jardin Botaniques,
      192 route de Lausanne, 1202 Genève, SWITZERLAND

Mr E W B H Milne-Redhead, Bear Lane, Colchester, Essex, ENGLAND

Mr A F Mitchell, Silviculturist, Forestry Commission, Alice Holt
      Research Station, Farnham, Surrey, ENGLAND

Mr R J Mitchell, Curator, University Botanic Garden, St Andrews,
      Fife KY16 8RT, SCOTLAND

Dr B A Molski, Director, The Botanic Garden of the Polish Academy
      of Science, 02-973 Warszawa 34 SKR 84, POLAND

Mr J K Muir, Curator, City of Liverpool Botanic Garden,
      Mansion House, Calderstone Park, Liverpool L18 6JL, ENGLAND

Madame Nicot, Museum National d'Histoire Naturelle, 57 Rue Cuvier,
      Paris 5, FRANCE

*Dr O E G Nilsson, 1st Curator, Botanic Garden, Villavägen 8,
      S75236 Uppsala, SWEDEN

Mr J Nurmi, Curator, Botanical Garden, University of Turku,
    SF-20500 Turku 50, FINLAND

Mr L A Pemberton, Supervisor of Studies, Royal Botanic Gardens,
    Kew, Richmond, Surrey TW9 3AB, ENGLAND

Mr V T H Parry, Chief Librarian & Archivist, Royal Botanic Gardens,
    Kew, Richmond, Surrey, TW9 3AE, ENGLAND

Mr A Paterson, Curator, Chelsea Physic Garden, 66 Royal Hospital
    Road, London SW3 4HS, ENGLAND

Dr F Perring, Biological Records Centre, Institute of Terrestrial
    Ecology, Monks Wood Experimental Station, Huntingdon, ENGLAND

Dr J Popenoe, Director, Fairchild Tropical Botanic Garden,
    10901 Old Cutler Road, Miami, Florida 33156, U.S.A.

*Mr H-H Poppendieck, Institut für Allgemeine Botanik und Botanischer
    Garten, D-2000 Hamburg, Heston 10, FEDERAL REPUBLIC OF GERMANY

Dr A Rannestad, Scientific Affairs Division NATO, 1110 Bruxelles,
    BELGIUM

*Dr P H Raven, Director, Missouri Botanical Garden,
    2315 Tower Grove Avenue, St Louis, Missouri 63110, U.S.A.

Mr A M Reguerio y Gonzalez-Barros, Conservador, Real Jardin Botanico,
    Plaza de Murillo 2, Madrid, SPAIN

Mr W Richter, Technical Manager, Neuer Botanischer Garten der
    Universität Göttingen, 34 Göttingen, An der Lutter 55,
    FEDERAL REPUBLIC OF GERMANY

Mr J B Rickell, Regional Grounds Maintenance Officer,
    Department of the Environment, Room 1120A, St Christopher
    House, Southwark Street, London SE1 0TE, ENGLAND

Professor Dr S Rivas-Martinez, Director, Real Jardin Botanico
    Plaza de Murillo 2, Madrid, SPAIN

Dr G Rossmann, Scientific Officer,
    Ruhr-Universität Bochum Botanischer Garten,
    4630 Bochum-Querenburg, Buscheystrasse 132, Postfach 2148,
    Bochum, FEDERAL REPUBLIC OF GERMANY

Professor M Ryberg, Director, Bergius Botanic Garden,
    (Hortus Botanicus Bergianus), 10405 Stockholm, SWEDEN

Mr A D Schilling, Deputy Curator, Wakehurst Place, Ardingly,
    Haywards Heath, Sussex, ENGLAND

Dr W Schultka, Institution Botanischer Garten der
    Justus-Liebig-Universität, 63 Giessen, Senckenberstrasse 6,
    FEDERAL REPUBLIC OF GERMANY.

Dr W Schultze-Motel, Curator, Botanical Garden and
    Botanical Museum, Koenigin-Luise-Strasse 6-8, D-1 Berlin 33,
    FEDERAL REPUBLIC OF GERMANY

*Mr R L Shaw, Curator, Royal Botanic Garden, Edinburgh EH3 5LR,
    SCOTLAND

Mr A A Sijuade, Curator of Botanic Garden, University of Ife,
    NIGERIA

*Mr J B Simmons, Curator, Royal Botanic Gardens, Kew, Richmond,
    Surrey TW9 3AB, ENGLAND

Mr C Sipkes, Tuinarchitect B.N.T., Rockanje, NETHERLANDS

*Mrs Magdalena Peña de Sousa, Orchid Curator, Jardin Botanico,
    Universidad Nacional Autonoma de Mexico, Mexico City 20 DF
    MEXICO

Dr W T Stearn, British Museum (Natural History), Cromwell Road,
    London SW7, ENGLAND

Professor E Steiner, Director, Matthaei Botanical Gardens,
    University of Michigan, Ann Arbor, 1800 Dixboro Road,
    Michigan 48104, U.S.A.

Dr A L Stork, Curator, Conservatoire Botanique, 1 Chemin de
    l'Impératrice, CH-1292 Chambésy, Genève, SWITZERLAND

Dr V Strgar, Curator, Biotechnicue Fakultete in Institue Builogiyo,
    6100 Ljubljana, Trouske 15, YUGOSLAVIA

Professor P Sunding, Director, Botanical Garden, Univeristy of Oslo,
    Trondheimsveien 23B, Oslo 5, NORWAY

Mr D Supthut, Director, Städtische Sukkulentensammlung,
    CH-8002 Zurich Mythenquai 88, Zurich, SWITZERLAND

Mr A M H Synge, Conservation Unit, Royal Botanic Gardens, Kew,
     Richmond, Surrey TW9 3AE, ENGLAND

Miss C Teune, Hortus Botanicus, University of Botanic Garden,
     Nonnensteeg 3, Leiden, NETHERLANDS

*Mr G Thomas, Gardens Consultant, National Trust, Briar Cottage,
     21 Kettlewell Close, Horsell, Woking, Surrey, ENGLAND

Dr S K-M Tim, Brooklyn Botanic Garden, 1000 Washington Avenue,
     Brooklyn, New York 11225, U.S.A.

Mr E Timbs, Administration, Royal Botanic Gardens, Kew, Richmond,
     Surrey TW9 3AB, ENGLAND

Mr H Townsend, Assistant Curator, Decorative Section,
     Royal Botanic Gardens, Kew, Richmond, Surrey TW9 3AB, ENGLAND

Mrs P D Turpin, Conservation Unit, Royal Botanic Gardens, Kew,
     Richmond, Surrey TW9 3AE, ENGLAND

Professor Dr W Vent, Director, Bereich Botanik und Arboretum des
     Museums für Naterkunde an der Humboldt-Universität zu Berlin,
     DDR 1195 Berlin - Baumschulenweg, Späthe-Strasse 80/81,
     GERMAN DEMOCRATIC REPUBLIC

Dr S M Walters, Director, University Botanic Garden, 1 Brookside,
     Cambridge, ENGLAND

Mr J F Warrington, Assistant Curator, Tropical Section,
     Royal Botanic Gardens, Kew, Richmond, Surrey TW9 3AB, ENGLAND

Professor D A Webb, Botanic Garden, Trinity College, Dublin 2,
     EIRE

Professor Dr H Webber, Director,
     Institut für Spezielle Botanik und Botanischer Garten,
     Universität Mainz, D-65 Mainz, FEDERAL REPUBLIC OF GERMANY

Mr D Wells, Nature Conservancy Council, Monks Wood,
     Experimental Station, Huntingdon, ENGLAND

Dr D O Wijnands, Scientific Officer, Hortus Botanicus,
     Plantage Middenlaan 2, Amsterdam, NETHERLANDS

Dr J T Williams, Department of Botany, University of Birmingham,
    Birmingham B15 2TT, ENGLAND

*Mr K R Woolliams, Horticulturist, Waimea Arboretum,
    59-864 Kamehameha Highway, Haleiwa, Hawaii 96712, U.S.A.

*Professor D Zohary, Department of Genetics, The Hebrew University,
    Jerusalem, ISRAEL

* Contributor of Paper

LIST OF ORGANISING STAFF - R.B.G. KEW

Internal Planning Committee

Mr R I Beyer
Dr P E Brandham
Mr J P M Brenan
Miss C A Brighton
Mr G Ll Lucas

Mr V T H Parry
Miss L M Ponsonby
Mr J B Simmons
Mr E Timbs

Exhibits

Miss D P Clarke
Mr C M Erskine
Mr B Halliwell
Mr S Henchie

Mr D Hunt
Mr P Reid
Mr J Ruddy

Reception

Miss A Chabert
Mrs D Emanuel
Mr A J Hale

Mrs S Kozdon
Mr E Timbs

Airport Reception

Dr P E Brandham
Miss C A Brighton
Mr C M Erskine
Miss M A Johnson

Mr J L S Keesing
Mrs P Turpin
Mr H Synge
Mr J Warrington

Photography and Projection

Mr T A Harwood
Mr L A Pemberton

Mr R R Zabeau

Discussion Recording

Miss Y Aspland
Mrs J Brace
Miss A Chabert
Mr C M Erskine
Mr H Fliegner
Mr D Field

Mr J L S Keesing
Miss S Leche
Dr S Owens
Mr H Synge
Mr D W H Townsend

Conference Dinner

Miss A Chabert

Mr A Harvey

Typing Pool

Mrs M Brind
Mrs J Ronald

Book Production Typing

Mrs E Attwood
Mrs J Curtis
Mrs E Fitchett

Tour of Kew

Mr G E Brown
Mr C M Erskine
Mr H Fliegner
Mr B Halliwell
Mr A Harvey

Mr J Mateer
Mr A D Schilling
Mr J Warrington
Mr J R Woodhams

LIST OF EXHIBITS AND EXHIBITORS

Mrs M.T.F. de Almeida
    Coimbra Botanic Garden,
    Portugal.

Photographs and examples of old
plant labels

Dr E. Ayensu
    Smithsonian Institution,
    Washington DC, U.S.A.

Coloured photographs of endangered
USA species

Professor T.W. Böcher
    University of Copenhagen,
    Denmark.

Photographs, Flora of Greenland,
and reprints

Mr B. Bruinsma
    Leiden Botanic Garden,
    Netherlands.

Booklets on Botanic Garden and
examples of old plant labels

Dr G. Ciuffi
    Florence Botanic Garden,
    Italy.

Photographs of Botanic Garden and
booklets on same

Professor Dr H. Demiriz
    Botanic Garden, University
    of Diyarbakir, Turkey.

Dried flowers, *Centaurea* and
Cynara from SE Anatolia

Professor Dr K. Esser
    University of the Ruhr,
    Federal German Republic.

Audio-visual aids and booklets on
Botanic Gardens

Dr C. Farron
    University of Basle,
    Switzerland.

SEM photographs. Necklaces made
from seeds and fruits

Professor C. Gomez-Campo
    University of Madrid,
    Spain.

Seed bank list of endemics of
Iberian Peninsula

Mr H.G. Hillier
    Hilliers Arboretum,
    Winchester, England.

Exhibit of cut material of
uncommon woody species

325

Dr M. Hofmann   Computer print-outs
    Buhlenweg,
    Federal German Republic.

Mr C.R. Lancaster   Information and photographs of
    Hilliers Arboretum,   Hilliers Arboretum
    Winchester, England.

Dr J. Miege   Posters and seed lists of
    Geneva Botanic Garden,   Geneva Botanic Garden
    Switzerland.

Mr C. Sipkes   Living plant material
    Rockanje,
    Netherlands.

Dr S. Tim   Photographs and booklets on
    Brooklyn Botanic Garden,   New York Botanic Garden
    New York, U.S.A.

Royal Botanic Garden   Computer print-outs
    Edinburgh,
    Scotland.

Royal Botanic Gardens   Conservation Exhibit. Plant
    Kew, Surrey,   records and labelling. British
    England.   native plants.

Dr M. Walters   East Anglian rare plants
    Cambridge Botanic Garden,
    England.
In conjunction with The Nature Conservancy Council

Mr K. Woolliams   Photographs of Hawaiian plants
    Waimea Arboretum,
    Hawaii, U.S.A.